THE LIVING CELL

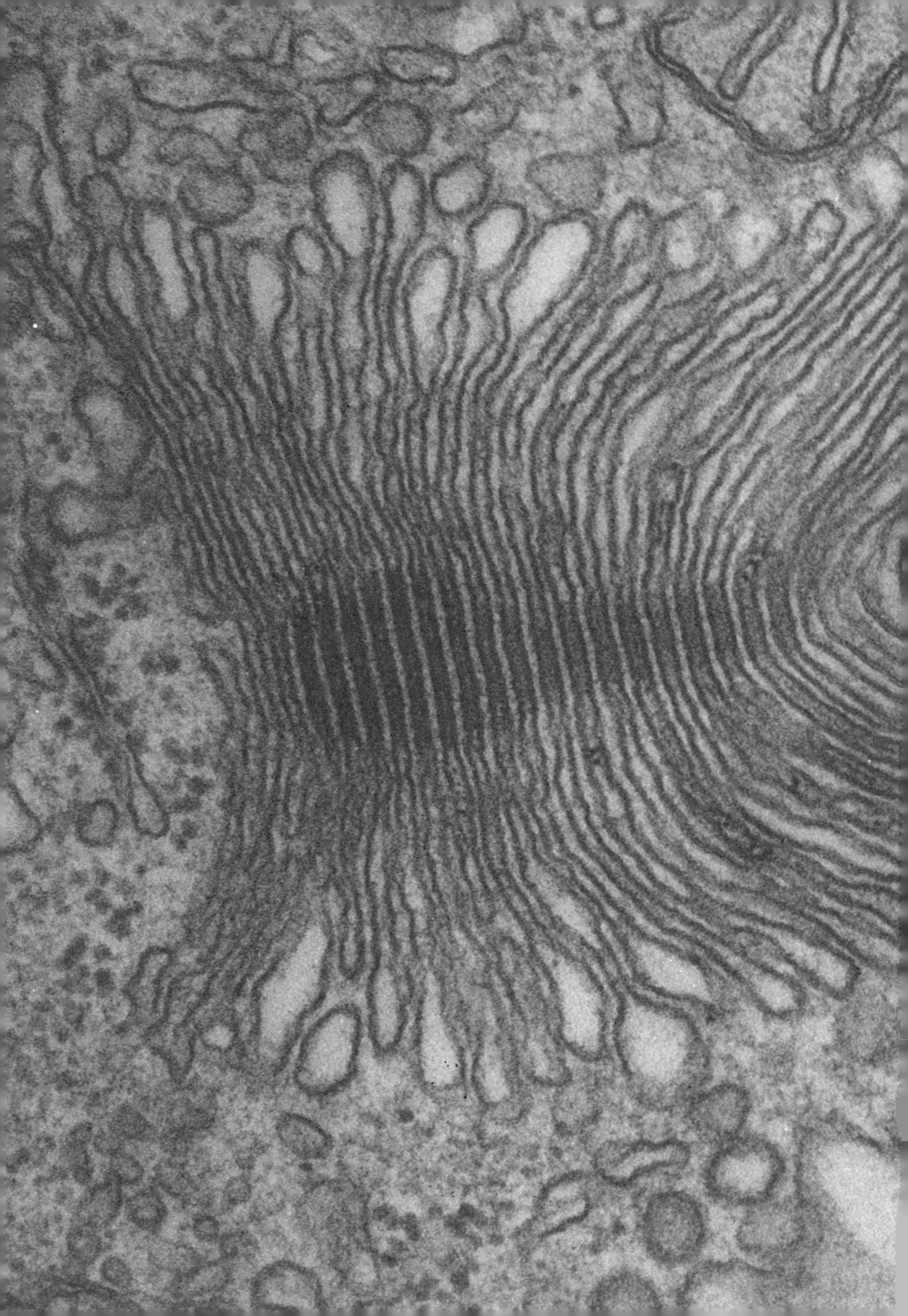

THE WORLD OF SCIENCE LIBRARY

GENERAL EDITOR: ROBIN CLARKE

THE LIVING CELL

Oliver Gillie

FUNK AND WAGNALLS NEW YORK

To my parents

Library of Congress Catalog Card Number: 76-119033

First American edition published in 1971 by arrangement with
Thames & Hudson International Ltd. (London)

Funk & Wagnalls, *A Division of* Reader's Digest Books, Inc.

Printed in the Netherlands

CONTENTS

PREFACE

The discovery that all living things are composed of cells modelled on a basic pattern has been more important than any other in giving biology a sure foundation. How cells work and exist in vastly different conditions in the bodies of animals and plants or as free living organisms is the subject of this book.

The progress of biological knowledge is often portrayed as an intellectual battle of scientists versus vitalists. The scientists are cast as the heroes with powers of super-thought and the vitalists as muddle-headed villains. I may have tended to follow this conventional picture but I believe a closer look at the heroes would show that they stand out because of their ability to cross this phoney intellectual barrier and take the best to be offered by different disciplines of thought. The molecular biologists are those who have done this most recently. Their approach has been so successful that it may not be long before they are able to manufacture artificial living cells. More than ever before biologists hold the power to influence and change human life as well as to give us a fundamental understanding of its basis.

I should like to acknowledge the debt I owe to Clifford Smith who first taught me biology and to Professors Michael Swann and C. H. Waddington and their colleagues too numerous to name who further stimulated my interest in biology at Edinburgh University. Lastly I should like to thank my parents for their longstanding encouragement and my wife, Louise Panton, for her understanding.

Oliver Gillie
London, 1970

WHAT IS LIFE? 1

There are so many differences between a frog and a stone that we do not hesitate to call the one living and the other inanimate. However, what is obvious in ordinary life is sometimes not at all obvious to science. Commonly held assumptions such as 'something is either alive or dead' may hide the most difficult problems – or so it seems to those who are prepared to worry about them.

A living thing may move or breathe, but not all living things do so. Plants do not move, and although most plants require in a sense to breathe, there are those, such as yeast, that can live without air. Living things reproduce themselves, but then so can non-living things such as crystals or even – as is described in chapter 5 – simple models consisting of mechanical wooden blocks. A crystal put into a saturated solution of the same substance acts as a seed and many more crystals are formed – in a sense the original crystal reproduces itself. A related process is the formation of the semi-precious stone called moss agate by the branching growth of a crystalline iron impurity in a quartz crystal.

A flame has many properties which living things also possess: it moves, it grows and it could even be said to reproduce itself; moreover, a flame can be said to exist at a point where energy is being used up to create something – in this case the fire and the heat

Opposite: a frog and a stone – the one alive, the other inanimate. Since both are composed of matter the difference between them must ultimately consist in the way their matter is organized. Many biologists now believe that life can be understood as a function of the complexity with which otherwise inanimate matter is organized in living creatures

Below: not a living thing, but a stone – a cut and polished moss agate. Its exquisitely patterned surface, vascular in appearance, is the result of crystalline 'growth'

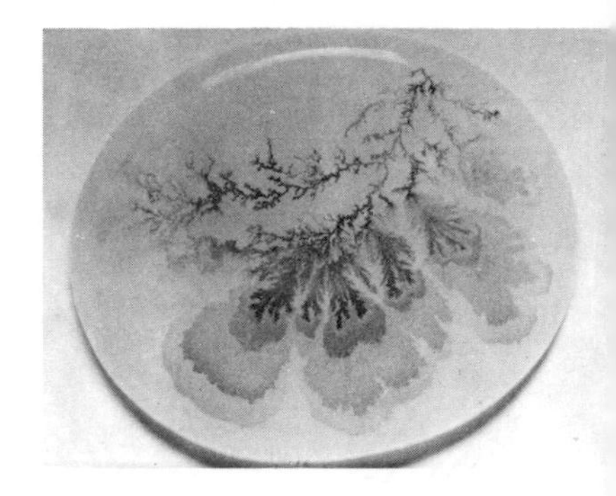

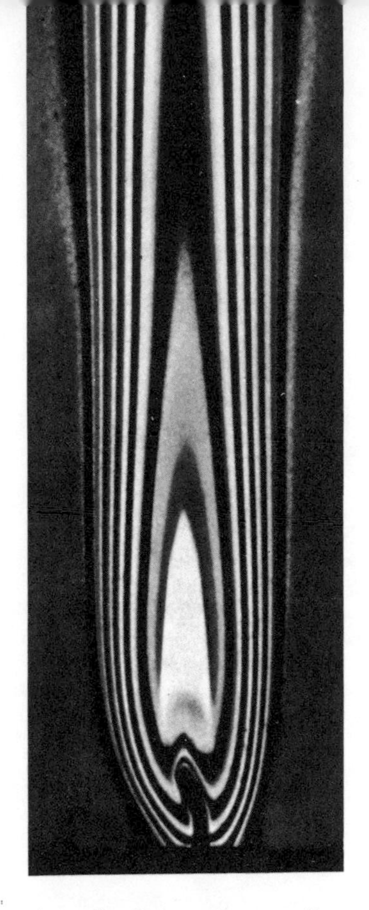

A candle flame and a spiralling silicon carbide crystal. Not only are both beautifully ordered structures, but the dynamism they exhibit is also in some respects analogous to that of living things.

around it. Living things are continually consuming energy: either the energy in the sun's rays or the energy stored by other organisms. And living things also give out warmth as a by-product of the energy which they use. A flame, however, is a flux which exists at a point where accumulated energy is being rapidly dissipated; it is unstable and disappears when the source of energy is consumed; whereas living things have evolved means of harnessing energy bit by bit so that it is dissipated slowly. They have also found ways of surviving periods when external sources of energy are not immediately available. Fire simply consumes the energy source and then dies.

Life and the cell

There are many basic similarities in the structure and organization of all living things which make them recognizable. The least complicated living things known have machinery which is so finely wrought that the most intricate man-made machines seem incredibly gross by comparison. The analogies which have been drawn between living cells and flames or crystals are really very crude – even though millions of years ago, when life was first evolving, there may have existed living, self-reproducing things only a stage more complicated than flames and crystals.

It is difficult to imagine how the large animals and plants most familiar to us could have evolved from simple chemicals. It is much less difficult, however, once we appreciate that animals and plants are made up of living cells which are all, without exception, organized in basically the same way. As well as the cells of which familiar animals and plants are composed, there are many microscopic one-celled creatures – bacteria, fungi, and protozoa such as the amoeba. These independent living cells also have the same basic organization as the cells of multicellular organisms. It is only by looking at the way in which living cells work, both individually and as part of multicellular organisms, that we can begin to understand what life is. A single cell is a very elaborate chemical machine,

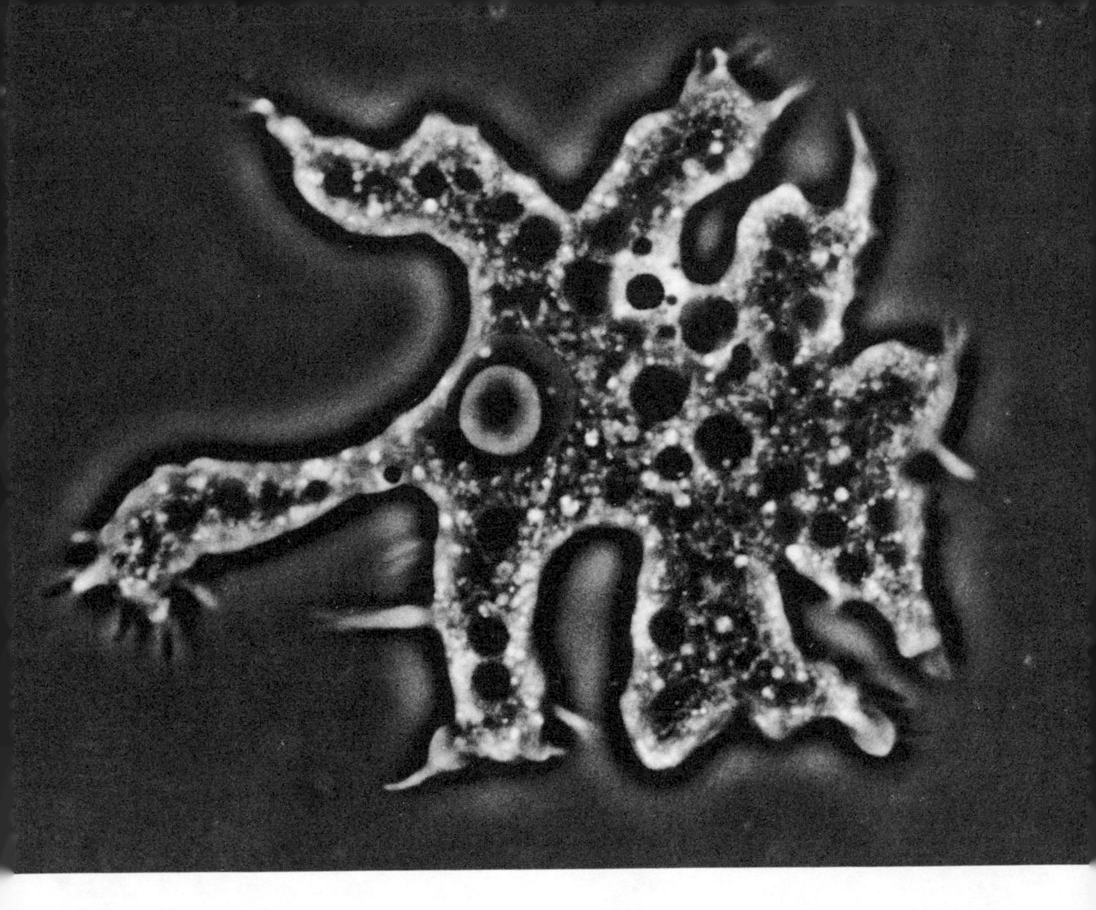

An amoeba. Although this tiny one-celled creature is perhaps less obviously ordered than flame or crystal, and over a thousand times smaller than the size shown here, it is in fact an enormously complex piece of chemical machinery possessing all the attributes of a living system

and so multicellular organisms, including man, could also be viewed as chemical machines – only with an extra dimension of complexity.

In the last ten years or so, thanks mainly to the work of molecular biologists, it has become possible to give an intellectually satisfying account of how the cell is constructed and how it works. The molecular biologists now know a great deal about some of the extremely complicated molecules involved in cell chemistry. They have also discovered how the hereditary instructions passed on from one cell to its progeny are encoded in the structure of one of these molecules – deoxyribonucleic acid (DNA). And they are beginning to accumulate detailed evidence in favour of the theory that life originally evolved from inorganic matter.

Modern biologists do tend, therefore, to think of living organisms as rather elaborate chemical machines. But while we may regard the bodies of others in this way, we have special internal knowledge of our own bodies – our personal machines; and this gives us a special interest in knowing how the machine works, what we can expect from it, and how we can make the best of it. What do vitamins, hormones and drugs do to the cell and the body? How do germs – bacteria and viruses – cause illnesses? How do cells defend themselves against infection? Why are some children born deformed? How do the cells of men and women differ? The answer to these and many other questions can be obtained only from a study of the cell itself.

Opposite: part of a giant walk-through model of the interior of a typical living cell. The brightly-coloured, stylized structures represent some of the cell's more important internal components (organelles)

Are we in the eyes of science 'merely' complex, well-ordered machines, as these cartoonist's drawings crudely suggest? In what sense, if any, living things are more than the sum of their component parts is a matter of controversy among biologists with a philosophic bent

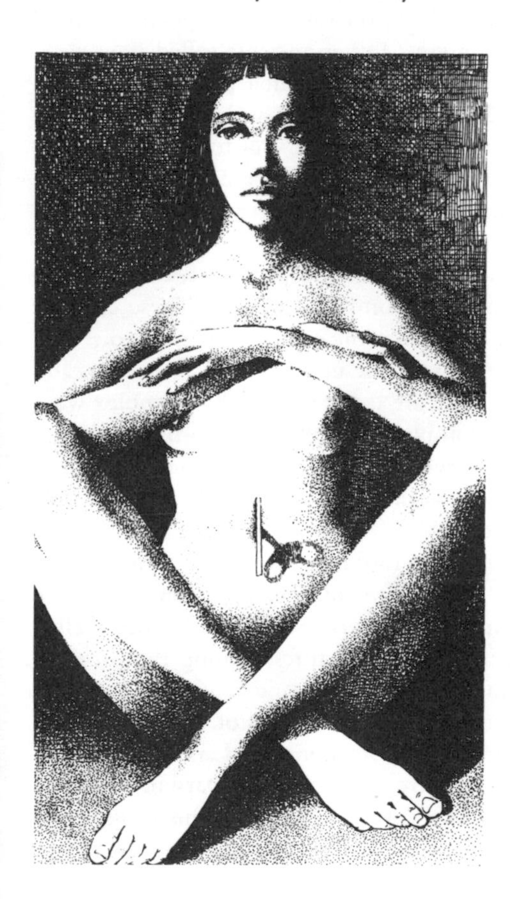

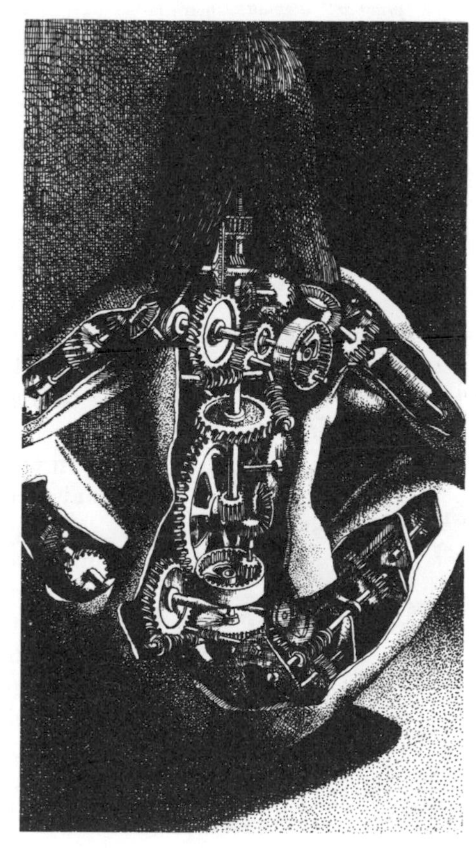

2 THE CELL AND ITS ORIGINS

A twelfth-century account of conception. A divine ordering element is shown descending into the womb of a pregnant woman, there to impart human form to a material seed

Many people still believe that some non-material vital force regulates living processes – a force to which laws of physics and chemistry cannot be applied. Aristotle called it the *psyche*. Other ancients called it *pneuma, anima* and *entelechia*. The French philosopher Henri Bergson, who died in 1941, believed in a creative force which he called the *élan vital*. But today such vitalist concepts – as they are called – are discounted by the scientist when he is being a scientist and not temporarily adopting a poet's viewpoint.

The bodies of animals and plants are, then, for the scientist, machines composed of cells. These cells are themselves little machines made up of molecules, which in turn are made up of atoms. The atoms and molecules of cells are arranged in highly complex ways – far more complex than in any non-living things. Nevertheless, cells have always, without exception, been found to work according to the laws of physics and chemistry.

The present extraordinary variety of living things must have arisen by evolution from one primitive organism millions of years ago in what has been called the primeval 'soup'. The fittest among the progeny of this living thing survived and reproduced. In this way a great many different kinds of organism arose, each well adapted to obtaining food and reproducing in a particular environment. Charles Darwin called this 'the struggle for existence'; and from this struggle has emerged the diversity of living things we know today.

Although many primitive organisms may originally have come into existence independently of one another, we can confidently assert that all the living things we are aware of must have been derived from a single original organism – because all living things are composed from basically similar types of cells. If cells built according to a different basic plan ever existed they must have become extinct in competition with the kind of cell that survives today. The basic similarity of cells is rather like the basic similarity of different types and makes of cars, lorries and motorcycles, which all have engines and wheels, but some use petrol, some diesel, and some electricity as a source of power, as well as differing in a great many other minor details.

Until the beginning of the nineteenth century all materials were classified as either organic (animal or vegetable) or inorganic (mineral), depending upon their origin. It was generally believed by scientists and others that organic materials could not be derived from inorganic materials. So it came as a great shock to the scientific world when in 1828 Friedrich Wöhler synthesized urea – a simple organic molecule found in urine – from inorganic chemicals. This was the first of a series of scientific discoveries showing that the living organic world was derived from the inorganic world. It was also the first step in the development of the science of organic chemistry, which has provided the essential basis for the growth of biological science, as well as being the basis of much of our industrial prosperity. The discoveries of modern biology are in a sense a repetition of Wöhler's demonstration that organic materials can be derived from inorganic – that the borderline between life and non-life is blurred.

Living matter

If a living material such as green wood is burned in a fire, after it is burned away only ash will be left. This ash is the mineral, or inorganic, residue of the wood, which is all that remains after burning. While the wood was burning carbon dioxide and water molecules were given off as gas and vapour. Molecules of water

MUCOPOLYSACCHARIDES
STARCH
UDP-N-Ac-Galactosamine
UDP-N-Ac-Muramate
UDP-N-Ac-Glucosamine
LACTOSE
Lactose-1-P
UDP-Galactose
UDP-Glucose
UDP-Glucuronate
UDP-Galacturonate
ASCORBATE
Glucose-1-P
GALACTOSE
Galactose-1-P
SUCROSE
GLUCOSE
MANNOSE
Mannose-6-P
Glucose-6-P
6-P-Gluconolactone
6-P-Gluconate
Glucosamine-6-P
Glutamine
Sorbitol
Fructose-6-P
FRUCTOSE
Fructose-1:6-di-P
Fructose-1-P
Erythrose-4-P
D-Xylulose-5-P
Sedoheptulose-P
D-Ribulose-5-P
D-Ribose-5-P
Ribulose-1:5-di-P
L-Xylulose
D-Xylulose
L-XYLOSE
L-LYXOSE
D-Arabitol
L-Arabitol
Arabinose
L-RIBULOSE
D-Xylose
D-RIBULOSE
D-ARABINOSE
D-RIBOSE
Ribitol
Glyceraldehyde-P
Dihydroxy-acetone-P
NAD-P
NAD-P 2H
1:3-di-P-Glycerate
3-P-Glycerate
2-P-Glycerate
P-enolpyruvate
PYRUVATE
LACTATE
ACETYL-CoA
GLYCEROL
Phosphatidate
Diglyceride
Triglyceride
FATS
FATTY ACID
ACYL-CoA
ETHANOL
Acetaldehyde
FORMATE
ACETATE
ACETONE
Acetoacetate
Acetoacetyl-CoA
Malonyl-CoA
Butyryl-CoA
Crotonyl-CoA
BUTANOL
CHOLESTEROL
Desmosterol
Zymosterol
Lanosterol
Squalene
Mevalonate
Isopentenyl-PP
Dimethylallyl-PP
Geranyl-PP
Farnesyl-PP
CAROTENOIDS
Geranyl-geranyl-PP
CO2
CITRATE
Cis-aconitate
Isocitrate
Oxalsuccinate
OXALACETATE
MALATE
FUMARATE
SUCCINATE
Succinyl-CoA
OXOGLUTARATE
Glyoxylate
ALANINE
CoA
Dephospho-CoA
PANTOTHENATE
Pantoate
4-P-Pantetheine
GLUTAMATE
Glutamine
LYSINE
Saccharopine
ASPARTATE
Asparagine
PROPANOL
PROPIONATE
Propionyl-CoA
Methylmalonyl-CoA
Citramalate
Mesaconate
P-Ribosyl-ATP
P-Ribosyl-AMP
P-Ribosyl-PP
Histidinol
HISTIDINE
HISTAMINE
Urocanate
Formiminoglutamate
AICAR
INOSINE-P (IMP)
XANTHOSINE-P (XMP)
GUANOSINE-P (GMP)
ADENOSINE-P (AMP)
Hypoxanthine
Xanthine
Urate
Allantoin
Allantoate
Urea
GDP
GTP
RNA
ATP
ADP
d-GDP
d-GTP
DNA
d-ATP
d-ADP
THYMIDINE-PP
d-TTP
d-TDP
d-TMP
Thymine
Deoxyuridine-P
d-UMP
d-CMP
d-CDP
d-CTP
CYTIDINE-PPP (CTP)
Uridine-P (UMP)
UTP
CDP
CMP
Cytosine
Uracil
Orotidine-P
Orotate
Dihydro-orotate
Dihydrouracil
NICOTINATE
FOLIC ACID C1 POOL
Formylglycinamide-RP
Glycinamide-ribosyl-P
Sarcosine
Dimethylglycine
Betaine
Glyoxylate
Oxalate
Glycollate
Glycolaldehyde
GLYCINE
Ethanolamine-P
ETHANOLAMINE
Phosphatidyl-serine
CEPHALIN
Phosphatidyl-ethanolamine
SERINE
Homocysteine
METHIONINE
S-Adenosyl-methionine
S-Adenosyl-homocysteine
CHOLINE
Choline-P
Cytidine-PP-choline
LECITHIN
Pyruvate
Hydroxypyruvate
Phosphoserine
Glycerate
CYSTEINE
CYSTINE
Cystathionine
Mercaptopyruvate
Cysteine sulphinate
Hypotaurine
Cysteate
Taurine
Shikimate
Shikimate-5-P
Chorismate
Prephenate
Anthranilate
TYROSINE
PHENYL ALANINE
Homogentisate
Dopa
Dopamine
Noradrenaline
Adrenaline
Phenylacetate
Benzoate
Melanin
THREONINE
Aminoacetone
Methylglyoxal
ISOLEUCINE
VALINE
LEUCINE
INDOLE
TRYPTOPHAN
Tryptamine
Indole pyruvate
Indoleacetate (Auxin)
Kynurenate
Kynurenine
Xanthurenate
Quinaldate
Picolinate
Homoserine
Aspartyl-P
Aspartyl-semialdehyde
Acrylyl-CoA
Malonic semialdehyde
Methylmalonyl-semialdehyde
Dipicolinate
Carbamyl-P
CITRULLINE
Arginino-succinate
ARGININE
ORNITHINE
UREA
Glycocyamine
Creatine
Creatinine
PROLINE
HYDROXY PROLINE
Putrescine
N-Ac-glutamate
N-Ac-Ornithine
Diaminopimelate
Pipecolate
PORPHYRINS
Porphobilinogen
Aminolevulinate
Glyoxylate
Pyruvate
LIGHT
ATP
Ferredoxin
Chlorophyll
Cytochrome
Plastoquinone
NAD(P)
NADH(P) + H+
NAD+(P)
FMN FAD
Ubiquinone
Cytochromes
Cytochrome oxidase
ADP + Pi
ATP
2H
O2
H2O
PHOTOSYNTHESIS
Redox Potentials (volts)
METABOLIC PATHWAYS
Carbohydrates
Aminoacids:
Biosynthesis
Degradation
Lipids
Purines and Pyrimidines
Biosynthesis
Degradation
Vitamins, Co-enzymes and Hormones
Photosynthesis
Miscellaneous reactions
A
B
C
D
E
F
G

consist of one atom of oxygen and two atoms of hydrogen. Molecules of carbon dioxide consist of one atom of carbon and two atoms of oxygen. The organic material of wood and other living substances consists of mineral ash, carbon, hydrogen and oxygen. Other elements such as nitrogen, sulphur and phosphorous, which may be contained in the mineral ash or given off as gases, are also important. These elements are the inorganic building blocks of the cells. They are joined together in various ways to form a vast inventory of organic molecules.

The cell is composed of four major types of organic materials: carbohydrates, familiar to us as paper, cellulose, starch (as in potatoes and flour) and sugar; fats, as found in butter, cooking oils and lard; proteins, as found in skin, leather, hair, wool and lean meat; and, lastly, nucleic acids, which are not familiar to us in ordinary life. There are also many other smaller organic molecules present in the cell which are destined to become parts of larger molecules. Carbohydrates and fats consist of carbon, hydrogen and oxygen; they are synthesized by plants using energy from the sun.

Opposite: A remarkable expression of the biochemical unity of life – a chart encompassing all the major chemical reactions and metabolic pathways involved in living processes. Yet it barely hints at the total molecular inventory and ultimate complexity of the living cell. (The chart is superimposed on an electron micrograph of multicellular vegetable tissue)

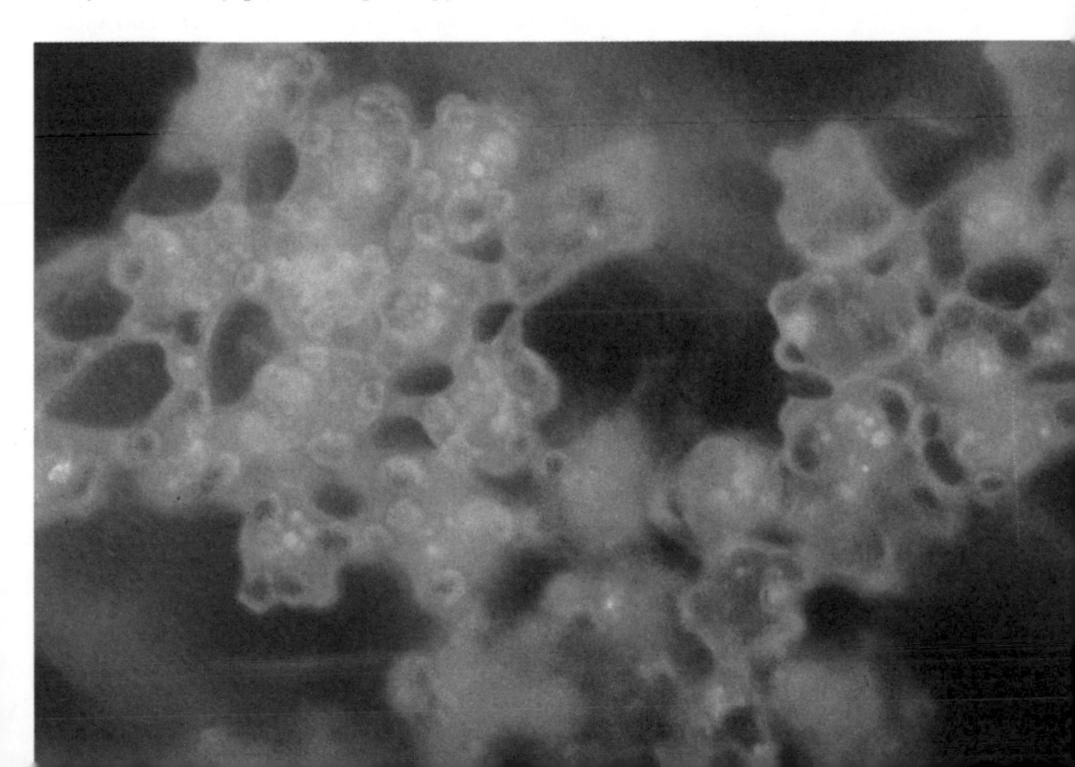

The organelles concerned with photosynthesis in green plants – the chloroplasts – show up red in this photomicrograph of leaf mesophyll tissue

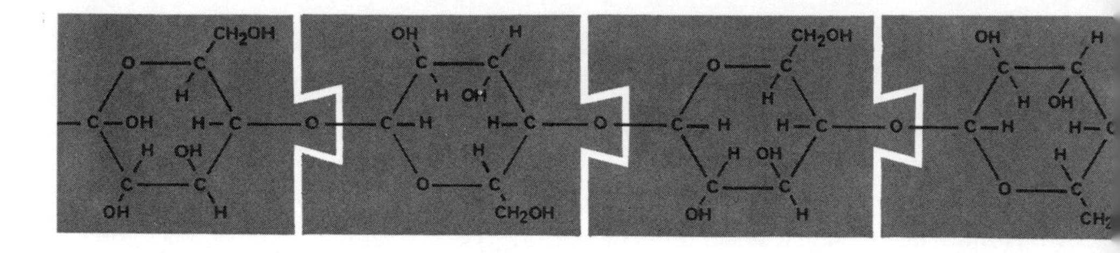

Cellulose molecules consist of large numbers – often many thousands – of glucose units linked end to end

They are used as an energy store, and also to make such parts of the cell as the fatty membranes and the cellulose walls of plant cells. Animals also synthesize carbohydrates and fats of their own, but they do this starting from carbohydrates and fats obtained from plants, or other animals, which they break down and resynthesize in their own way.

Proteins are made up from the elements carbon, hydrogen, oxygen and nitrogen, and sometimes sulphur. These elements are built up into molecules called amino acids, which in turn are joined together to make a protein molecule. There are only twenty different kinds of amino acids, but many thousands of different proteins. Each protein consists of a string of amino acids arranged in a special order. This special order gives each protein its individuality. There are two kinds of animal protein – structural and soluble. Structural proteins are found in hair, wool and skin, and also as gelatin in bones and cartilage. They have long fibrous molecules, which link together and give strength to the tissues they form.

Each thread in the much magnified cellulose membrane (below) is a bundle of many hundreds of cellulose molecules ranged in parallel

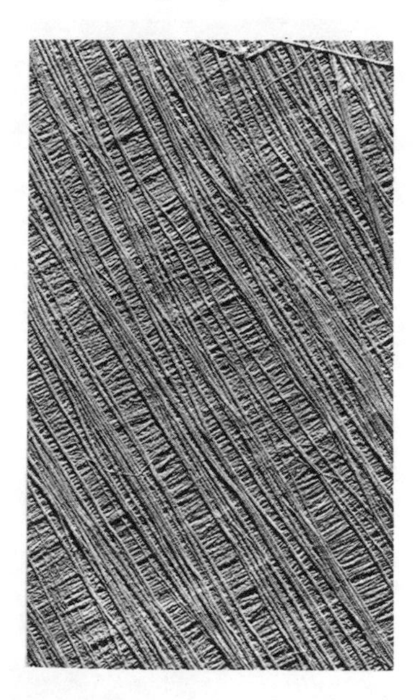

The soluble proteins are found inside the cell where they do its chemical work. These proteins are rounded globular molecules called enzymes and they speed up the various chemical reactions of the cell by joining together some molecules and splitting others apart. The enzymes work in interconnected pathways so that the product of one enzyme reaction is used by another enzyme to make yet another substance. These biochemical pathways are called the metabolic pathways of the cell or, simply, the cell's metabolism.

The fourth major material of the cell, the nucleic acid, was almost completely ignored until the 1940s and 1950s. During this period it was established that DNA (deoxyribonucleic acid) was the material which stored coded information for making proteins and passed it on from one generation to the next. DNA was shown by experiment to be the genetic, or hereditary, material of bacteria and viruses; and there is no doubt that it is also the genetic material of all living cells. (A few viruses are exceptional in having the related substance RNA – ribonucleic acid – as their genetic material.) The identification of DNA as the genetic material was perhaps the greatest blow to vitalism since Darwin published *The Origin of Species* in 1859.

DNA has been called the 'thread of life' and it has taken the central place in our understanding of the cell. It is a highly stable molecule consisting of four different kinds of units. The order of these units spells out a message, written in code, telling the machinery of the cell how to arrange the amino acids in the proteins. The DNA is a blueprint for the activities of the cell. The other nucleic acid, RNA, assists the DNA by carrying the genetic message bit by bit from the DNA to the cell's machinery for protein synthesis. Although the details of these processes are very complicated, in essence they are very simple. The DNA contains a message which is translated into protein, and in this way all the different proteins necessary to allow the cell to maintain itself and grow are specified. When the cell divides, the DNA message is transcribed exactly into two DNA copies of the original molecule, one of these copies going to each of the daughter cells during cell division. This process results in each daughter cell's having an exact copy of the DNA of the parent cell.

Even the simplest known cells utilize about fifty different proteins as well as DNA and RNA of various kinds. Such a structure is far too complicated to have come into existence all at once by chance, and so the first organisms must have been simpler arrangements of molecules. But would such organisms have consisted of protein or nucleic acid? In living cells proteins act as catalysts and nucleic acid acts as a stable self-

Right: a model of a short stretch of DNA in which each coloured sphere represents a single atom. To model a whole molecule millions of such spheres might be needed. The size of the DNA molecule – which may vary in different organisms – reflects the vast quantity of genetic information encoded along its two intertwined helices

copying information carrier – both seem to be equally indispensable. Which, then, came first: nucleic acid or proteins? This is the riddle of the chicken and the egg reduced to a scientifically intelligible question. Even so, it is hardly any easier to answer.

How did life begin?

Most geologists now believe that the earth came into existence some 5,000 million years ago, when countless small cold particles coalesced. The oldest fossil organisms that we can recognize are about 2,000 million years old. These organisms, which are found in Pre-Cambrian rocks, resemble colonial algae which still live today. There may be traces of much earlier living things in certain rocks, but how are we to recognize them unless they bear some resemblance to known living things? Chemical analyses of rocks as much as 3,100 million years old have shown the presence, in very small amounts, of organic molecules which might have originated from living things. However, it is

Embedded in rock more than 2,000 million years old, this minute fossil alga from Canada is counted among the earliest forms of life known to science

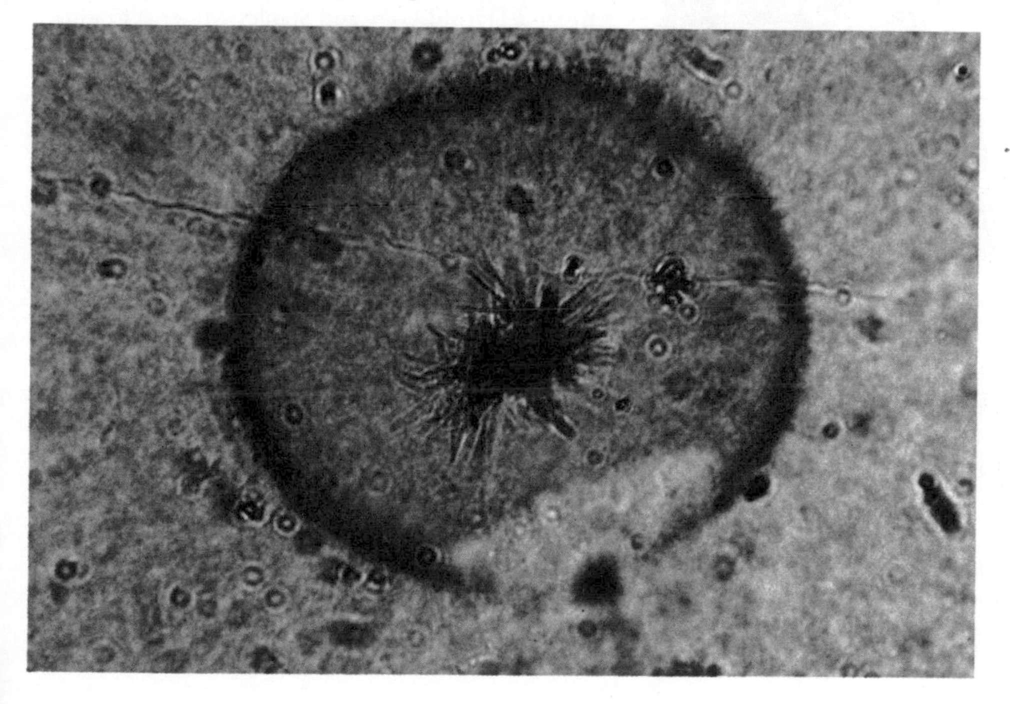

always difficult to be certain that this organic material did not merely percolate into the rocks at a later date or arise as a result of spontaneous chemical processes.

The synthesis of simple molecules of the kind found in living things by means of reactions possible on the primitive earth are beginning to show how life might have evolved. There is a lot of evidence suggesting that the atmosphere of the primitive earth was quite different from the present atmosphere. All the oxygen was probably combined in metallic oxides and in water molecules, and all the carbon was present either in elemental form or in metallic carbides. With an atmosphere consisting of hydrogen, methane, ammonia, hydrogen sulphide and hydrogen cyanide, but containing no oxygen, any organic molecules formed by some chance reaction would have been much more stable than they could possibly be under present conditions. But what are these chance reactions which might have led to the formation of the first organic molecules?

In 1952 Harold Urey and Stanley Miller, a graduate student in Urey's laboratories, made a bold and simple experimental approach to this problem. They mixed together water vapour and the gases methane, ammonia and hydrogen – all of which are believed to have been present in the primitive atmosphere. They circulated this mixture of gases in a glass apparatus, repeatedly passing through it an electric spark simulating lightning. Analysis of the contents of the glass apparatus after a week showed that the amino acids (constituents of proteins) glycine, alanine and aspartic acid had been formed in surprisingly large quantities.

This type of experimental approach has been extended by others and it has since been found that fourteen different amino acids and also some small proteins can be produced when hydrogen cyanide, ammonia and water are subjected to electric sparks in a closed vessel. This suggests that proteins may have been synthesized directly from simple gases in the primitive atmosphere. In fact, reactions have now been worked out which result in the synthesis – under the

Miller's experiment, carried out at the suggestion of Harold Urey. Water vapour (1) and the gases methane, ammonia and hydrogen (2) were circulated past an electric discharge (3) and through a cooling jacket (4). After a week a variety of newly synthesized organic molecules were found to have accumulated in the trap (5)

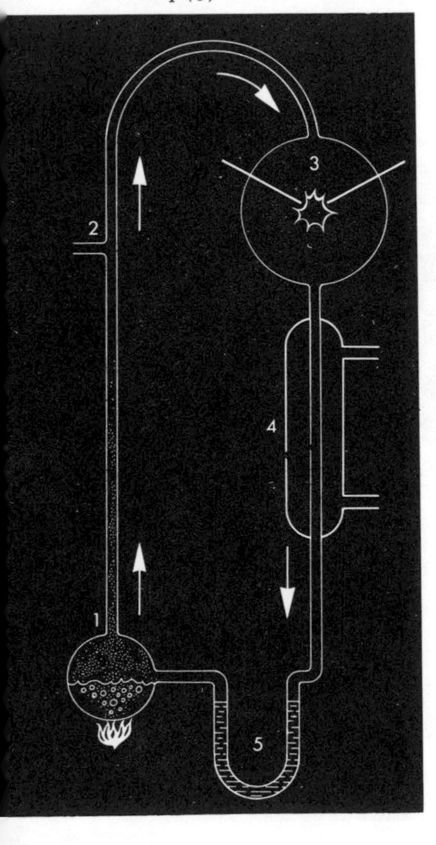

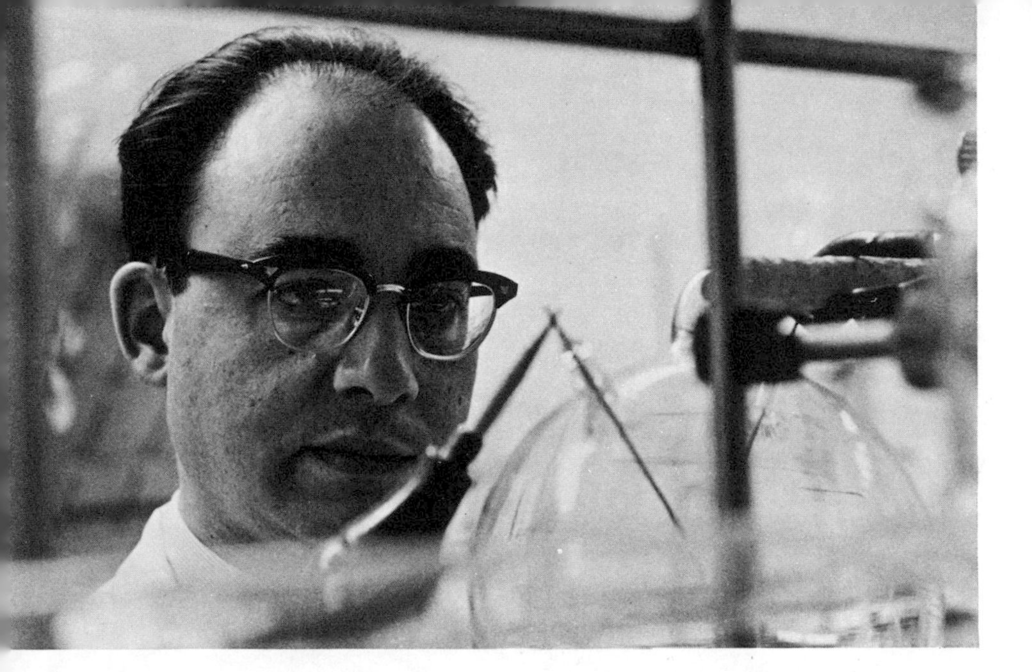

Stanley Miller with the apparatus for the famous 'life-synthesis' experiment in which he attempted to simulate the earth's primordial atmosphere

conditions of the primitive atmosphere – of all twenty amino acids found in the protein of living organisms today, and also of four of the five constituents of the two nucleic acids.

These reactions may have occurred in the atmosphere of the primitive earth for as long as a thousand million years before any living thing was made. But some time during that period at least one improbable event occurred. Perhaps it occurred several or even many times – no one knows. The improbable event was the coming together of certain molecules, synthesized in the atmosphere, into a form which was able to reproduce itself. Although the sea might have been a dilute soup of organic molecules it was probably too dilute for molecules to come sufficiently close together to form the first living things. More hospitable conditions would have existed in shallow rock pools, where evaporation would result in increasing concentrations of molecules. The first primitive living things might then have been washed into the oceans where, to begin with, they would find plenty of food awaiting them. At present it is not possible to say much about how these primitive organisms worked,

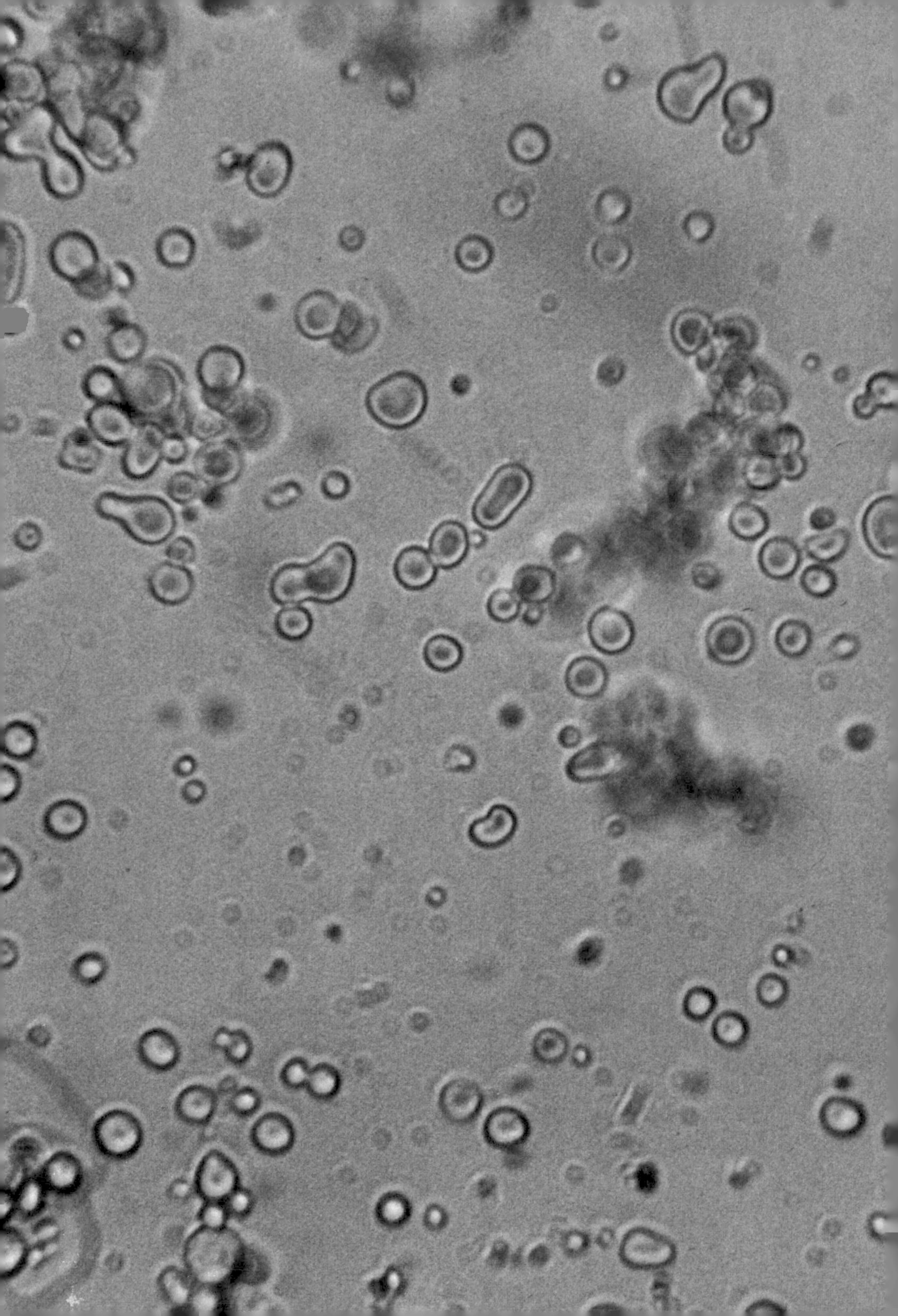

but they probably relied on mineral substances in their environment (instead of enzymes) to act as catalysts, and they must have grown very slowly. The first living things may have consisted entirely of nucleic acids, which could possibly have acted as both self-reproducing molecules and as catalysts – but this is only speculation. Once the first self-reproducing living thing came into existence, however, it would be expected to evolve rapidly as a result of natural selection of the 'fittest' types.

Some scientists believe that the origin of life was not such an improbable event and that given our primitive atmosphere living things were an inevitable consequence of inevitable chemical reactions. Professor Sidney Fox has pointed out that molecules have a capacity to order themselves which may have led to the spontaneous appearance of life by processes involving a gradual increase in order rather than by some very remote chance event. This view has considerable appeal and Professor Fox has adduced experimental evidence to support it.

He has found that if amino acids are heated together, simple proteins are formed which contain a selection of the amino acids present in the original mixture. These simple proteins are called proteinoids. In water or in salt solutions the proteinoids were found to

These microspheres (opposite) which form spontaneously when artificially synthesized proteins are dropped into water of the right acidity, are suggestively cell-like. In particular, each has a double wall surrounding it and, also, if the acidity of the water is increased, may 'reproduce' by splitting in two. Could tiny globules like these possibly have been the precursors of living cells?

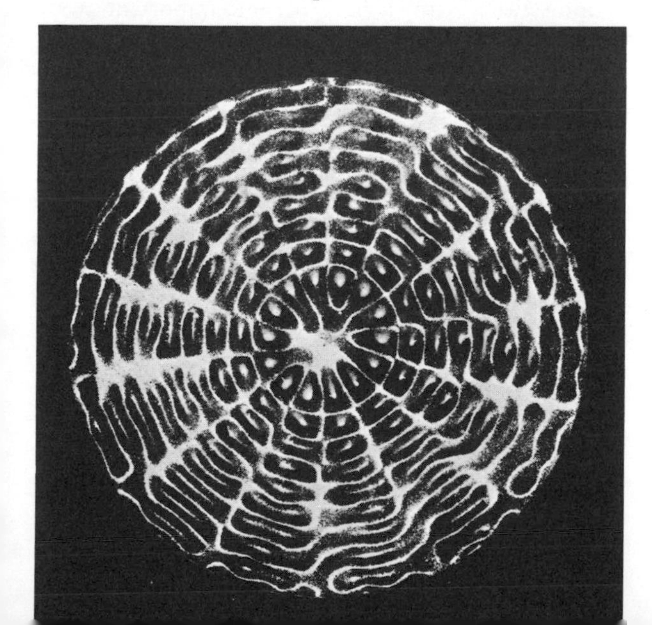

In another suggestive experiment a powdered metallic oxide scattered evenly over a vibrating diaphragm forms a distinctive and unexpectedly complex pattern. Could the forerunners of cells have been selectively shaped by patterns of forces comparable in their rhythms and variety to musical compositions?

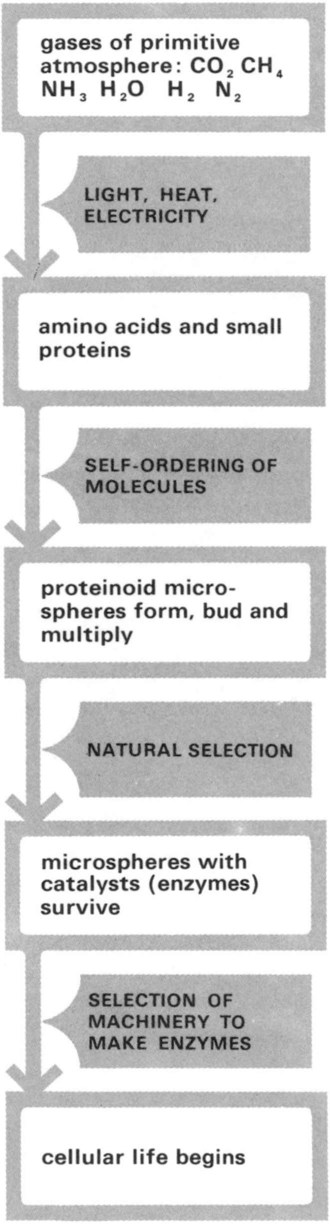

gases of primitive atmosphere: CO_2 CH_4 NH_3 H_2O H_2 N_2

LIGHT, HEAT, ELECTRICITY

amino acids and small proteins

SELF-ORDERING OF MOLECULES

proteinoid microspheres form, bud and multiply

NATURAL SELECTION

microspheres with catalysts (enzymes) survive

SELECTION OF MACHINERY TO MAKE ENZYMES

cellular life begins

assemble themselves into microspheres between 0·5 and 3 microns in diameter (1 micron equals one thousandth of a millimetre). This is about the same size as the smallest known living organism – the mycoplasma – and indeed these proteinoid spheres superficially resemble mycoplasmas. But perhaps the most remarkable thing about the microspheres is that they can grow (by accretion) and bud to give off daughter microspheres which grow in the same way. Such microspheres, formed in drying out pools thousands of millions of years ago, could have provided a sheltered place in which simple nucleic-acid molecules could reproduce themselves (but suggestions of this kind, which can never be more than speculative, however plausible they may appear, should be treated with some reserve).

The first living things had to survive in the presence of the primitive atmosphere, which contained no oxygen. So they would only be able to obtain energy from sugars by anaerobic fermentation, as yeast and other organisms do today. Whether or not sugars were the first energy source available to organisms is not known; energy might somehow have been obtained by direct absorption of heat into minerals or from chemical reactions involving mineral substances. A system for the utilization of energy from the sun (photosynthesis) must have become important at an early stage, as photosynthesizing organisms would have such great advantages over other organisms. Once photosynthesis began, carbon dioxide and water from the surroundings would be converted into sugar and oxygen, causing a complete change in the primitive environment. As a result of this, free oxygen would for the first time be liberated into the atmosphere. Organic molecules are much less stable in the presence of oxygen, but apparently some of the primitive organisms were sufficiently well organized to survive the change. The presence of oxygen in the atmosphere permitted the evolution of systems of biochemical reactions in which sugar is decomposed completely into carbon dioxide and water with a much greater release of energy. And this must have led to the evolution of the aerobic type of energy release commonly found in present-day organisms. But all this could not have happened in a simple proteinoid microsphere: machinery for accurate protein synthesis from inherited instructions must already have evolved. These deductions are not mere guesswork – there is a solid basis of experimental evidence on which they have been built, and it seems certain that experimentation will lead to increasingly detailed theories in the future.

What of the possibility of life on other worlds? If life were to arise on another planet under conditions identical to those of the primeval earth, would it develop in the same general way again? Or would it be quite different. Opinions differ, but chemists point out that life based on quite different kinds of chemistry is theoretically possible and may, indeed, have occurred in the early history of the earth itself. Some of these other forms of life may some day be encountered on other planets if space exploration is ever extended to other star systems.

Opposite: a scientist's tentative attempt – much simplified – to see the origin of life as the possible outcome of a sequence of natural events

Below: an illustration from the 1663 Cambridge Bible depicting the traditional, theological view of life as having its origin in a supernatural act of creation

CELL STRUCTURE 3

'Fibrous threads', *'rete mirabile'* (the wonderful network), 'gruel', 'slime' and 'noble juices' are some of the words that scientists of the sixteenth and seventeenth centuries used to describe living material. Robert Hooke, in his book *Micrographia,* published in 1665, was the first actually to describe cells. He examined a sliver of cork under his microscope and observed 'microscopical pores' – the spaces enclosed by the cell walls of dead cork tissue. But the general significance of this discovery was not appreciated, and Hooke's contemporaries thought that cells were just holes in a network of threads.

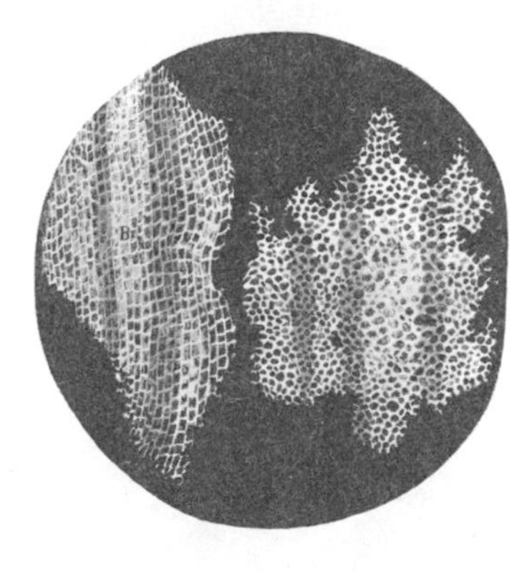

Above: the first visual record of the cellular structure of living matter – sketched by Robert Hooke in 1665. Opposite: what Hooke saw – a transverse section through vegetable tissue; only the cell walls remain. To some early observers such cell cavities appeared to be no more than holes in a network of threads

For some time cells were thought to exist only in plant material. But in 1838, during a dinner conversation between two German scientists, Matthias Schleiden and Theodor Schwann, a general cell theory was born. Schwann realized that plant cells described by Schleiden were very similar to structures he had observed in the nervous tissue of animals, and in due course he was to show that all tissues – blood, skin, bone, nerve and muscle – are composed of cells. He was also the first to point out that eggs were cells despite their comparatively large size.

The cell theory was a great intellectual achievement. The recognition that animals and plants were built from similar basic units was the first step towards the development of evolutionary theories and the idea that life had a single origin. And the cell theory was

Matthias Schleiden (top) and Theodor Schwann, co-founders of modern cell theory. Right: Schwann's drawings of plant and animal cells, intended to illustrate their fundamental similarity

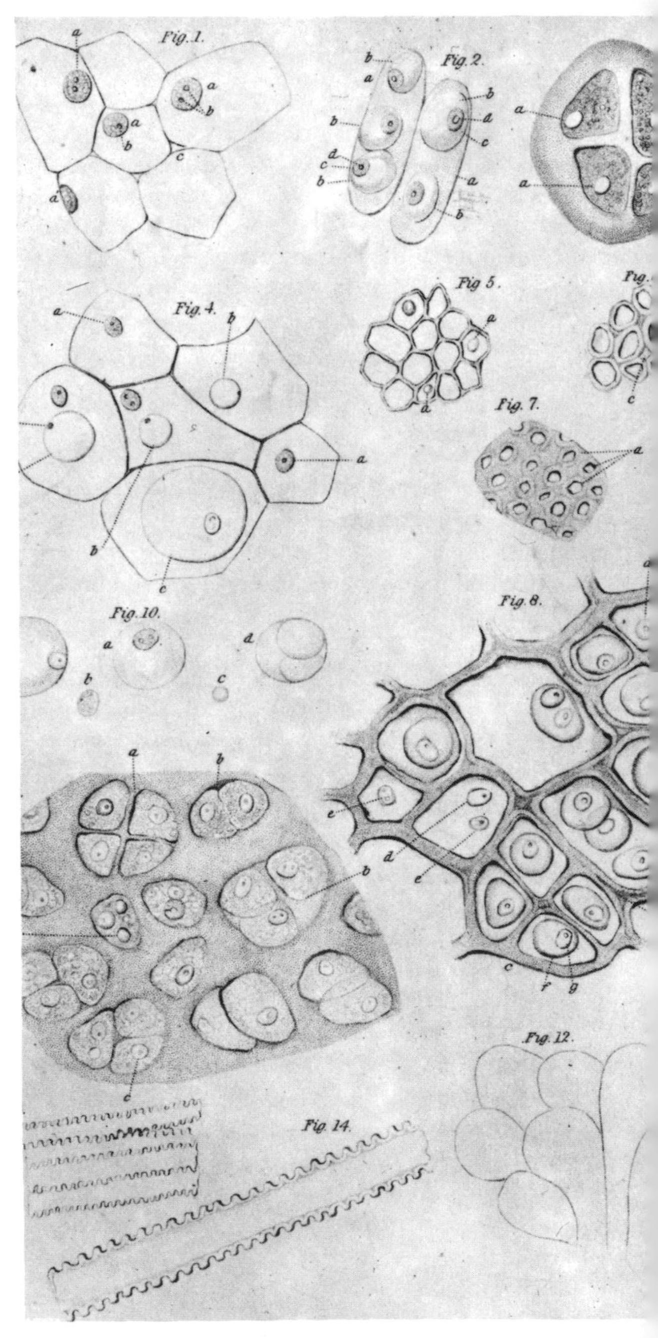

also essential to an understanding of embryological development and of the laws of inheritance.

As microscopes improved, evidence for cell theory accumulated, as more and more structures could be discerned within the jelly-like substance of the cell. Jan Evangelista Purkinje, an outstanding Czech anatomist who had had a religious training, called the jelly of the cell 'protoplasm' after the theological term 'protoplast', which means the first formed being. The protoplasm was at first thought to be just liquid, and later to be a colloid, like paste or clay. But eventually particles and fibrils could be distinguished within it. The protoplasm has since proved to be composed of an enormous variety of remarkable structures. Many of these structures have been intensively studied under the light microscope, but it has taken the resolving power of the electron microscope to reveal the organization of the cell in significant detail.

Distinguishing the organelles

Professor E. Ruska built the first electron microscope in 1933. Since then these instruments have been steadily improved and can now magnify one million times: points no more than one thousandth of a micron apart can now be distinguished. An electron beam (equivalent to the light beam in a light microscope) passes through a specimen on to a fluorescent screen, or photographic plate, where electron intensity is converted into light intensity, giving us a picture. Dark areas of this picture represent parts of the specimen which are dense to electrons and light areas of the picture represent parts of the specimen which are transparent to electrons. Most biological materials are rather transparent to electrons, so in order to ensure adequate contrast, suitable stains are used in the same way as stains are used in light microscopy. These stains are all heavy metal compounds, which are very dense to electrons. The heavy metals are taken up to a different extent by different parts of the cell so revealing cellular structure.

The great disadvantage of the electron microscope

Microscope used by Hooke. Specimens to be examined were fixed on a large pin and illumination was provided by an oil lamp, the light being condensed by a water-filled glass sphere and focused by a small convex lens

is that it cannot be used to observe living material, because the specimen has to be held dry in a vacuum to avoid disturbing the electron beam. For this reason it is not possible to understand how the cell really works simply by looking at it – even if one uses an electron microscope.

It is the test tube, rather than the microscope, which has made the cell understandable in terms of the atoms and molecules of which it is composed. Nineteenth-century chemists analysed and synthesized organic compounds and found that they were made from the same chemical elements as inorganic compounds. They isolated and identified many of the most important chemical constituents of living cells – including haemoglobin, the red pigment of blood; and chlorophyll, the green pigment in plants. They also studied the roles of enzymes in the processes of digestion.

Shortly after the turn of the century the approaches of biochemists and cell biologists began to converge; though it has only been during the last twenty years, with the growth of molecular biology, that cell biology has really flourished. The usual approach of the biochemist is to break open cells by grinding them up with sand, or by homogenizing them by some mechanical means. He then studies the chemistry of the parts of the cell. This approach may seem exceedingly crude, rather as if someone hoped to deduce how a motor car worked by knocking one to pieces and studying the broken parts. It is of course possible to come to quite the wrong conclusions using this sort of method: the observation that the tyres of a car will dissolve in the petrol is of no help in understanding the workings of a car and might be positively misleading. Nevertheless, it is in avoiding such pitfalls that the skill of the biochemist lies. Once the biochemist has ground up his cells he usually proceeds to separate out different parts of the cell by some process of purification. The method most frequently used is to spin test tubes full of homogenized cells in a centrifuge for varying lengths of time; the heaviest parts of the cell are soon thrown to the bottom of the tube and the

Two key scientific instruments in the study of the living cell. Opposite: an electron microscope capable of magnifying specimens up to a million times. Below: an ultra-centrifuge powerful enough to subject cell tissue to forces of up to half a million times that of gravity

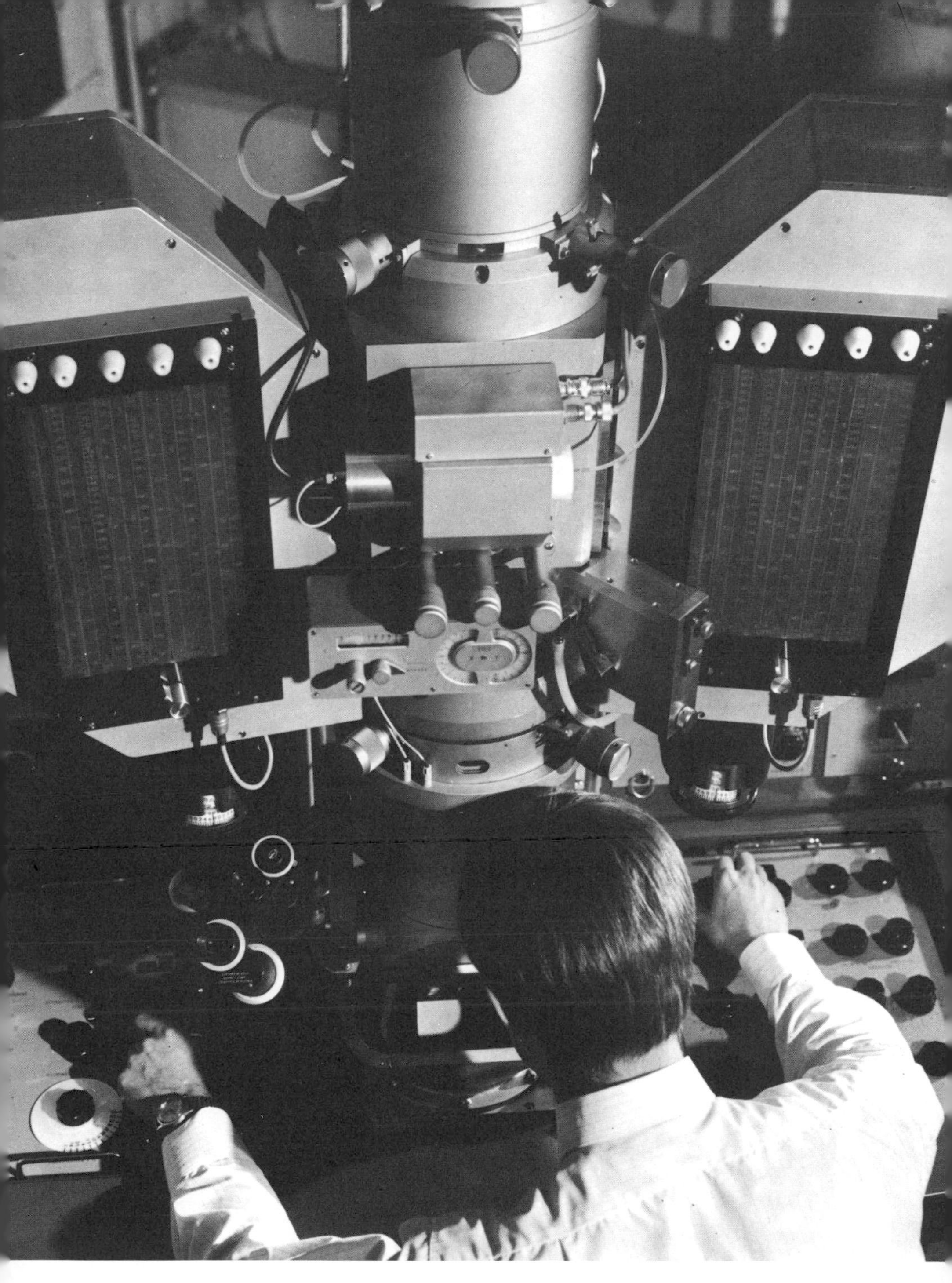

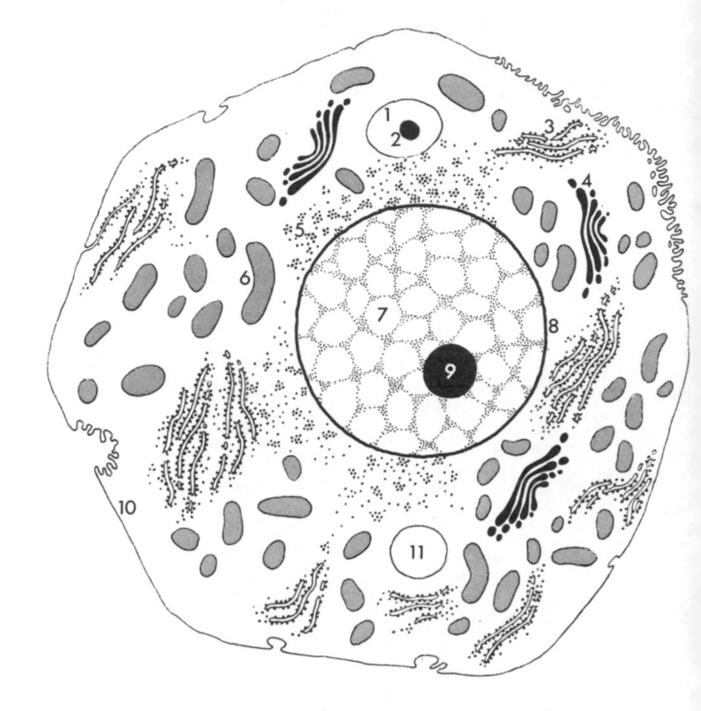

centrosome 1
centriole 2
endoplasmic reticulum 3
golgi apparatus 4
ribosomes 5
mitochondrion 6
nucleus 7
nuclear membrane 8
nucleolus 9
cell membrane 10
vacuole 11

Above: Diagram of a generalized animal cell, locating the main organelles. Right: electron micrograph of an animal cell, magnified about 5,000 times, in which most of the organelles included in the diagram can be identified

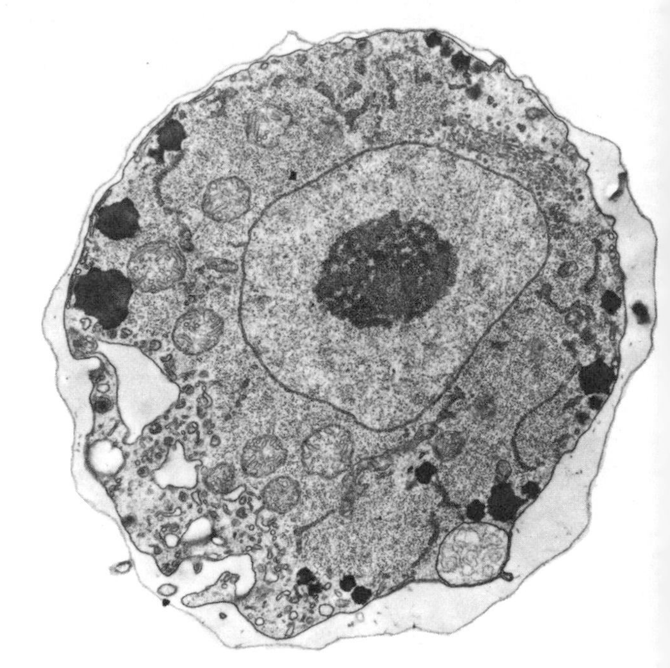

lighter particles are thrown down progressively until there remains only the cell sap, which contains all the soluble substances of the cell including the enzymes.

The cell nucleus, containing the cell's quota of DNA, is the first part of the broken cell to sediment when centrifuged. Next come the various cell organelles – little structures performing specialized tasks within the cell. Among these are: the mitochondria – the energy-providing 'batteries' of the cell; the green chloroplasts (found only in plant cells), which make sugar, using energy from the sun; granules of starch – the cell's food stores; the membranes that synthesize the proteins of the cell; and the lysosomes – the 'suicide bags' – which contain enzymes powerful enough to destroy the cell itself.

All these organelles can be separated from one another and purified by using powerful centrifuges. The first high-speed centrifuges, invented by a Swede, Theodor Svedberg, in 1925, were able to generate a force hundreds of thousands of times as great as that of gravity, and attain speeds of more than a million

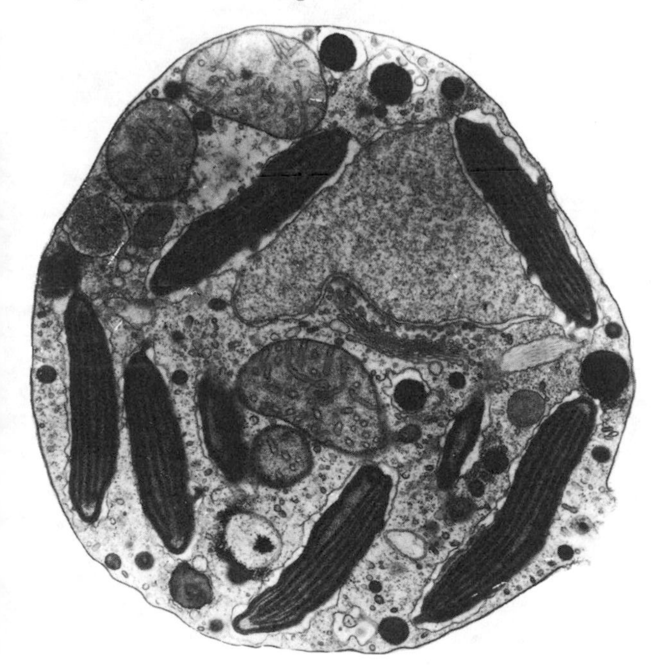

A plant cell (left) differs most conspicuously from an animal cell in that in addition to the usual organelles it is equipped with chloroplasts (the prominent longitudinally-striped bodies) enabling it to utilize solar energy in photosynthesis. And instead of a flexible cell membrane it has a rigid cellulose cell wall. Here the magnification is 10,000 times

revolutions per minute. The introduction of these high-powered instruments to biochemistry in increasing numbers since the Second World War has brought about a technical revolution perhaps even greater than that occasioned by the electron microscope.

Science proceeds by gathering information in whatever way it can and then by putting this information together to get an overall perspective. The studies of the light microscopists, electron microscopists and biochemists thus complement each other to give a composite picture of the cell and its parts. This picture has been put together piece by piece, rather like the pieces of a jigsaw puzzle, and although there may still be some pieces missing we can now see the general plan. All cells are found to possess membranes, ribosomes, a nucleus of some kind and various organelles, which may include mitochondria, chloroplasts, lysosomes and other less common structures.

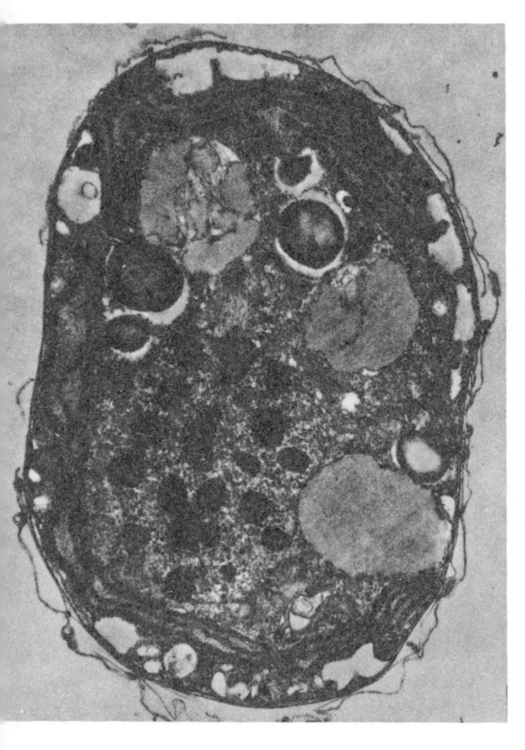

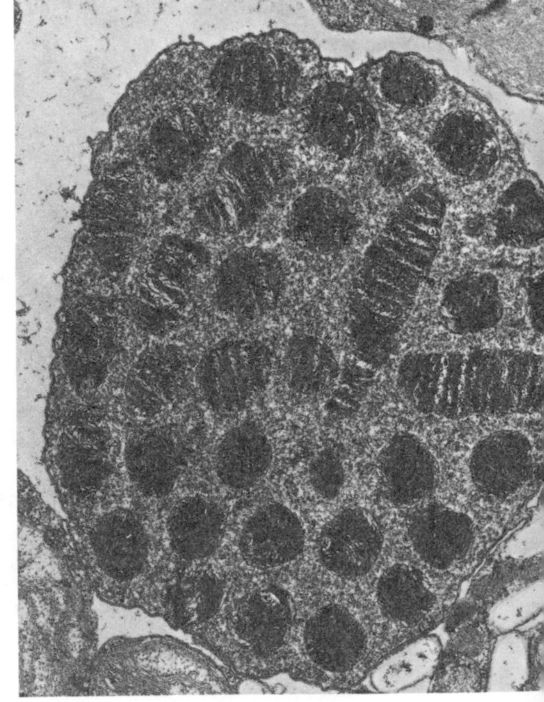

The nucleus

If the nucleus of a cell is carefully removed with a fine needle, the cell may still go on living in the sense that it is using up energy and oxygen; but even the hardiest of cells cannot live more than a week or two without a nucleus. The nucleus is the cell's command centre, and a cell without a nucleus, like a body without a head, must die. The nucleus of a non-dividing cell seems to have little structure when viewed under a microscope. However, if cells are stained with specific stains DNA, RNA, and protein can all be shown to be present in the nucleus. The chromosomes themselves are visible only in the dividing cell and can be seen to contain all the DNA of the nucleus. The nucleus is surrounded by a membrane which is much the same as the other membranes of the cell – it is pierced by holes providing a connection between the cell sap and the nuclear sap.

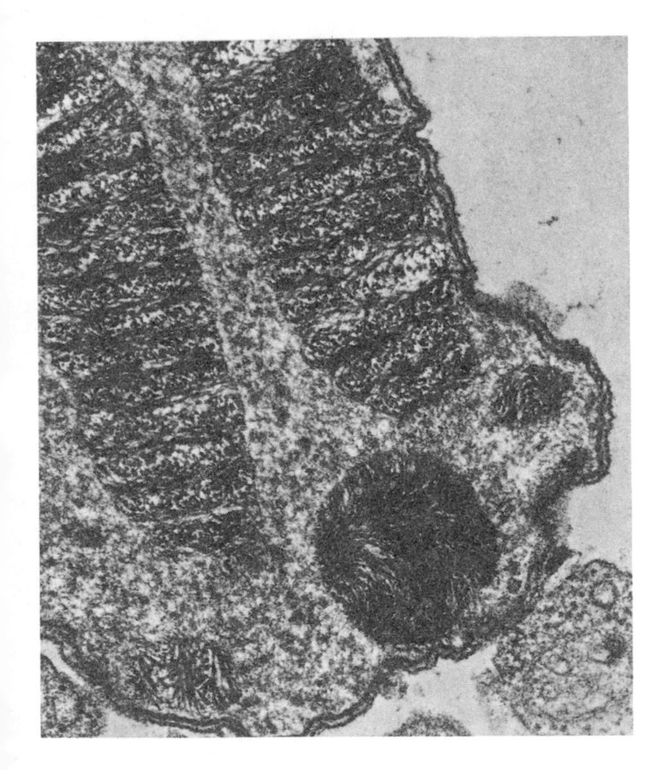

Sections through the nucleus of a single-celled animal, successively enlarged from left to right. The prominent circular and elongated bodies within the nuclear membrane are chromosomes, seen in transverse and longitudinal section. The chromosomes in the third micrograph are magnified 48,000 times and it is possible to distinguish the fine fibrils of DNA of which they are composed

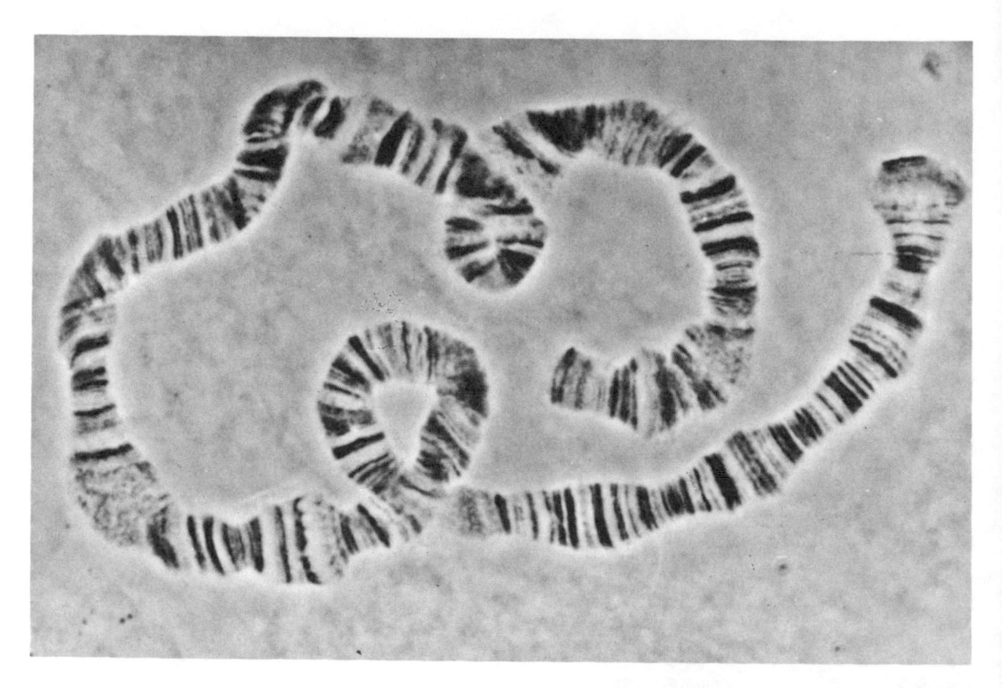

A clearly banded chromosome from the salivary gland of the fruit fly Drosophila. The darker bands contain a higher proportion of DNA than the lighter bands, which consist mostly of RNA and protein

The structure of chromosomes was for a long time a great puzzle and has only become clear in the last few years. Besides the DNA in the chromosomes, there is a great deal of special protein called histone, which was for a long time considered to be the important constituent of chromosomes, while the DNA was ignored. In 1903 Walter S. Sutton and Theodor Boveri suggested that the hereditary particles, or genes, were located on the chromosomes; and in 1913 A. H. Sturtevant suggested that the genes might be organized in a linear sequence along the chromosome. We now know that these genes are coded in the linear structure of the DNA molecule itself, and that the histone protein which covers the DNA plays some secondary role, protecting the DNA, or perhaps influencing the expression of the genes in some way. It was one of the great achievements of genetics to have deduced from experiments involving the breeding of animals and plants that heritable factors were located in a linear order on the chromosome before the linear structure of DNA had ever been suggested.

Chromosomes may be studied most easily in certain special cells in which they are unusually large. The best known example is the giant chromosomes of the salivary-gland cells of the fruit-fly *Drosophila*. They contain a thousand times as much DNA as normal chromosomes and are about three hundred times as long. These chromosomes can be seen to have a characteristic pattern of bands along their length which is constant from fly to fly. The DNA appears to be present in greater quantity in the banded parts than it is between the bands. Experiments have been made to discover the structure of these chromosomes by breaking them down with pure digestive enzymes which specifically destroy either RNA, DNA or protein. When the chromosome is attacked by enzymes which break down RNA or protein nothing much happens, but when it is attacked by enzymes which break down DNA the chromosome disintegrates. This is evidence that the integrity of the DNA strands must be maintained if the overall structure of the chromosome is to be preserved. The salivary-gland chromosomes consist of many identical strands of DNA, which lie roughly parallel between the bands but are coiled up in dense loops in the bands themselves. Normal chromosomes, however, have only one DNA strand, which coils up when the chromosomes become short and dense during cell division, uncoiling again after cell division, the chromosomes being no longer microscopically visible.

The bands on the salivary-gland chromosomes of the fruit-fly have been extensively studied by geneticists and it has been found possible in many cases to locate the exact position of a gene on the chromosome by microscopic examination. The salivary-gland chromosomes lie very close together, side by side, so that there appears to be only one chromosome rather than two. However, as in other cells the chromosomes are paired – one chromosome being inherited from the mother and the other from the father. If one of these parental chromosomes is deficient or damaged at any point some bands will be missing on just one of the

A mutation, or abnormal change in a chromosome's gene structure, often produces a conspicuous change in external appearance. Here, the 'bar-eye' mutation in Drosophila is illustrated. Instead of its normal round shape the eye is narrowed to a bar

two paired chromosomes. An abnormality of this kind, called a deficiency, can easily be seen under the light microscope.

If flies which are known to have such a deficiency are bred together, the offspring which inherit the deficiency in both of their chromosomes will usually die. But it appears that the cells can work perfectly well if they have only one copy of a gene, or piece of chromosome, instead of two (i.e. a single deficiency). They cannot survive if they have no copies of the gene at all, however. A double deficiency of an identical piece of chromosome almost always proves lethal to the organism. In the converse situation, where a cell inherits an extra piece of chromosome called a duplication, the effect need not be at all harmful, even when present in both chromosomes – and may sometimes be advantageous. Extra genetic material duplicated like this in the chromosome is almost certainly of great evolutionary importance to the organism. The genes in the duplicated material can mutate, and so, perhaps, evolve, without disturbing the existing organization of the cell. There is one particular duplication of the genetic material in *Drosophila* called 'bar' eye, which causes the eye to become slit shaped when present in both chromosomes.

The salivary-gland chromosomes of insects are also of special interest because they show 'puffs' at various characteristic places along their length. These puffs are places where the DNA strands have spread out and are involved in synthesizing a lot of RNA. There is a characteristic change in the puffing pattern of the chromosomes as the insect develops from a grub into an adult fly, showing that different genes are being activated as various processes of development occur. The pattern of puffing also occurs in a characteristic way in different tissues of the developing organism. If a grub is injected with moulting hormone (a hormone which induces an insect to shed its skin and develop a new and larger one), particular parts of the chromosome are excited into activity and produce puffs. This is a startling demonstration that the genes not only contain information and instructions for making the organism but also respond to the influences of the cellular environment in which they find themselves.

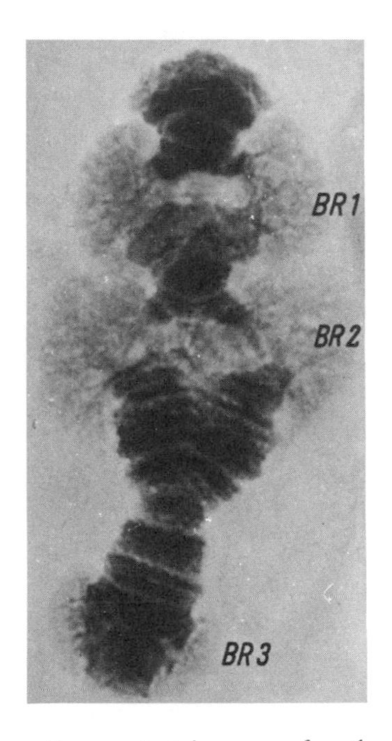

Above: a giant chromosome from the salivary gland of the midge Chironomus tentans, with three very prominent puffs. Below: an autoradiograph of the same chromosome injected with radioactive uridine. The spots indicate intensive manufacture of RNA coded to produce saliva proteins

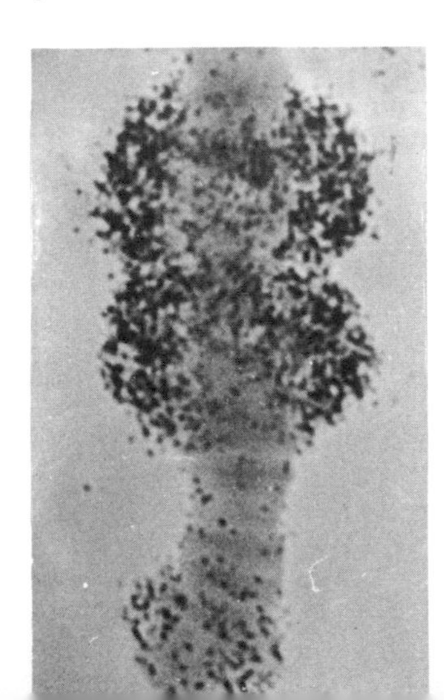

Most of the RNA in the nucleus is found in one or more little bodies called nucleoli, which are visible only in those cells which are not in the process of dividing. The nucleoli are parts of the chromosome where large quantities of RNA are being synthesized to make the RNA of the ribosomes. A genetic mutant of the frog is known which has only one nucleolus instead of two in every cell. When two of these frogs are mated 25 per cent of the offspring have two nucleoli per cell, 50 per cent have one nucleolus per cell and 25 per cent have no nucleoli at all (this is a typical pattern for the inheritance of genes according to Mendel's laws, and is discussed in the next chapter). The embryos that have no nucleoli are able to grow up into tadpoles because the egg contains enough ribosomes for them to develop that far. However, they die before they become frogs, because their cells run out of ribosomes.

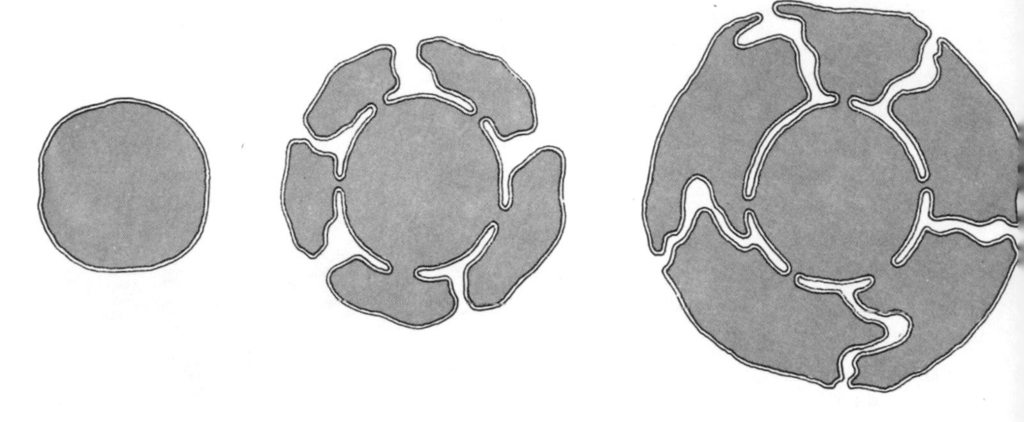

A much simplified diagram suggesting how, in principle, the living cell's complement of organelles might have evolved by infolding of the original cell membrane

The cytoplasm

The nucleus and the cell organelles appear to float in the jelly called protoplasm, or more commonly nowadays, cytoplasm. Under the light microscope the cytoplasm appears to have little or no structure. Under the electron microscope, however, it can be seen to consist of layer upon layer of membranes rather like the skins of an onion. These membranes, called the endoplasmic reticulum, all have the same basic structure and they are all continuous with one another. In fact, the nuclear membrane is continuous with the membranes of the endoplasmic reticulum as well as with the membranes of the Golgi apparatus – another organelle in the cytoplasm – and these membranes are in turn continuous with the outer membrane of the cell. All the membranes of the cell can be thought of as having evolved from an original outer membrane which has proliferated to provide a greatly increased 'surface' and become sufficiently convoluted to divide the cell into compartments where special activities can be carried on separately.

Two types of membrane, called 'rough' and 'smooth', are found in the endoplasmic reticulum. The rough membranes, unlike the smooth, are dotted with particles, called ribosomes, which are involved in protein synthesis. The ribosomes consist of protein and

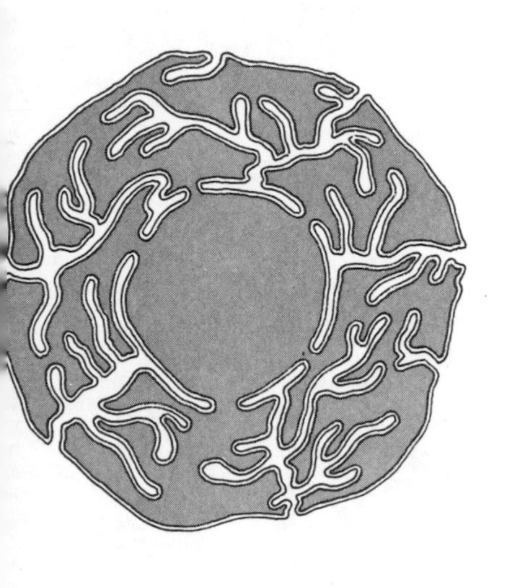

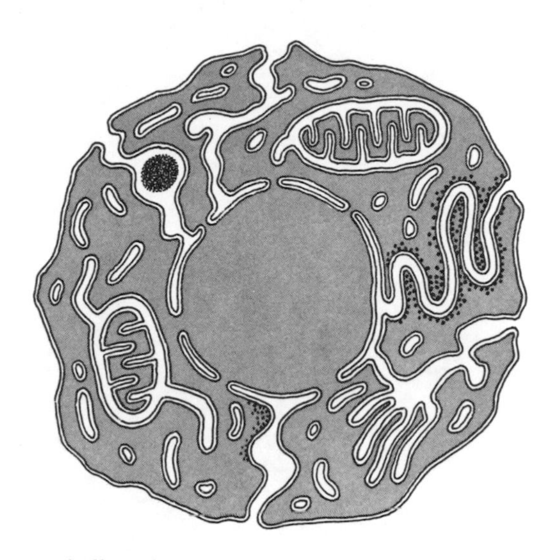

RNA in about equal quantities, and, as we shall see in greater detail in chapter 5, they combine with genetic messenger-RNA molecules to form polysomes which do the actual work of synthesizing the protein. Polysomes may comprise from six to thirty ribosomes, depending on the type of protein being synthesized. The ribosomes themselves are about 15 millionths of a millimetre in diameter, which is too small for much structural detail to be visible even under the electron microscope. However, the ribosomes can be separated into two unequal-sized parts by various treatments followed by centrifugation. The function of the ribosomes seems to be to hold the messenger-RNA molecule firmly while the transfer-RNA molecule reads out the message and helps to translate it into protein. The messenger molecules might be thought of as the printer's block, the ribosomes as the press, and the finished protein as the printed page.

The ground substance of the cell, which is found everywhere between the membranes, contains soluble proteins (enzymes of various kinds) and RNA. Also in the ground substance are structures called microtubules and microfilaments, which can sometimes be seen in electron microscope pictures. The function of these structures is unknown, but it has been suggested that

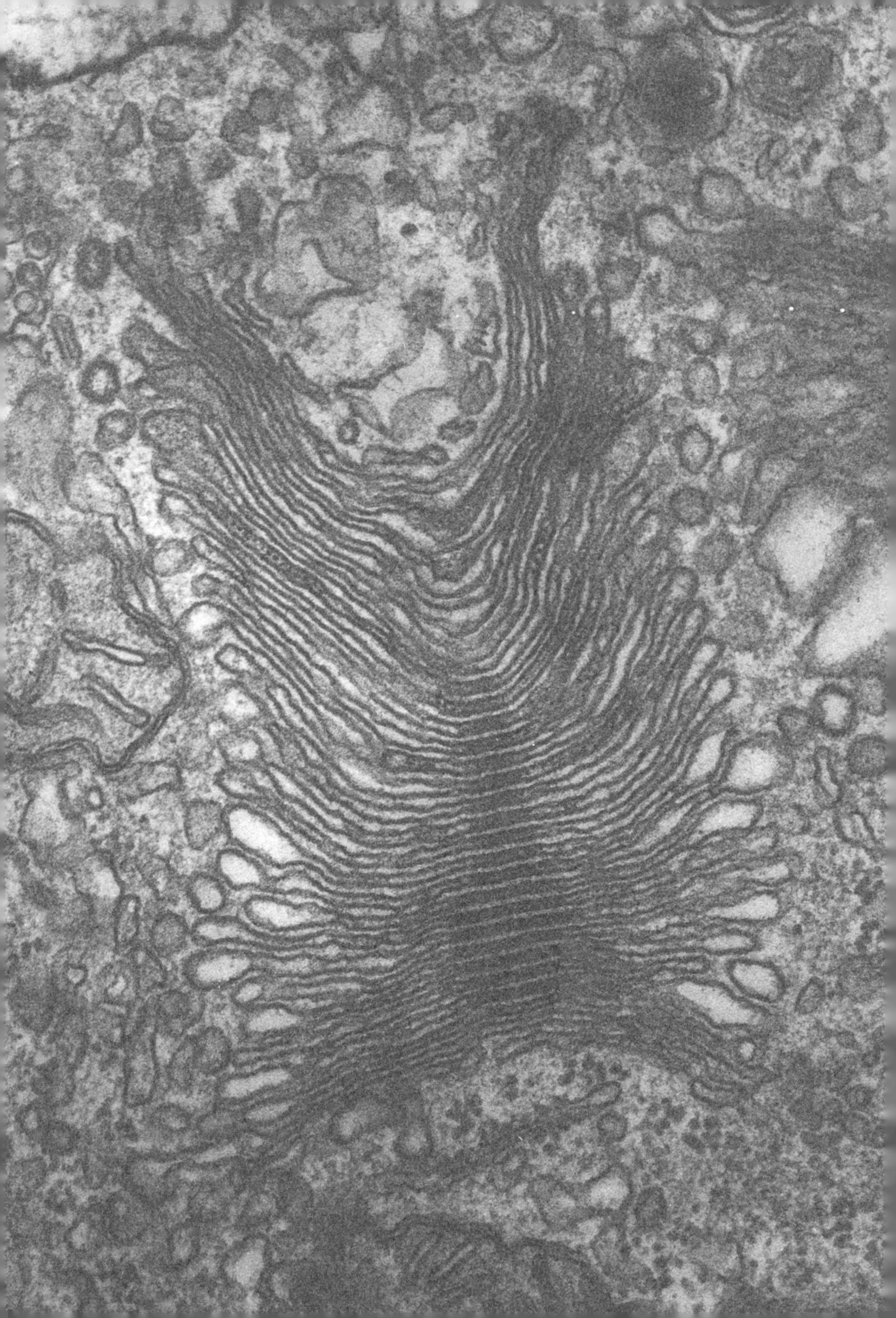

they may play a part as contractile fibres at cell division, provide mechanical support for the cell, or play a role in the transport of ions and molecules within the cell.

In 1898 an Italian, Camillo Golgi, made preparations of the nerve cells of the barn owl and the cat, which he stained with silver salts and osmium tetroxide. He saw a dark-staining part of the cytoplasm of the cell quite distinct from the nucleus. This structure, called after him the Golgi apparatus (it is called the dictysome in plants), is a system of membranes similar to the smooth membranes of the endoplasmic reticulum. For a long time its existence was disputed by biologists – partly because of its variable appearance in cells – but now electron microscope studies have established its existence beyond doubt.

Recently the Golgi apparatus has been shown to play a most important part in preparing and packaging proteins secreted by cells. Such proteins are made on the ribosomes in the normal way and then accumulate in the Golgi apparatus. The Golgi apparatus then adds some carbohydrate molecules to the protein molecule, and wraps together a large number of the molecules in a single membrane. This package may then pass to the edge of the cell and release its contents outside: the mucus part of saliva is made in this way in the cells of the salivary gland. And the digestive enzymes of the gut are also put into packets by the Golgi apparatus before being liberated into the intestine. However, the Golgi apparatus is also responsible for packing certain enzymes into parcels – called lysosomes – which remain within the cell.

The membranes of the cell – the nuclear membrane,

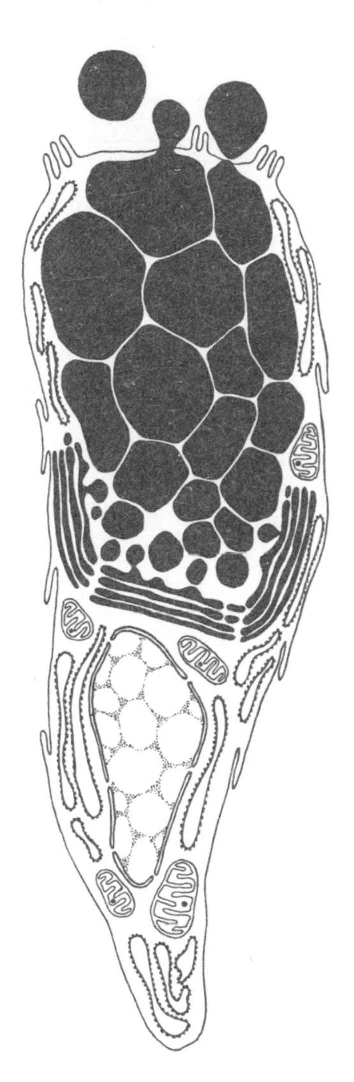

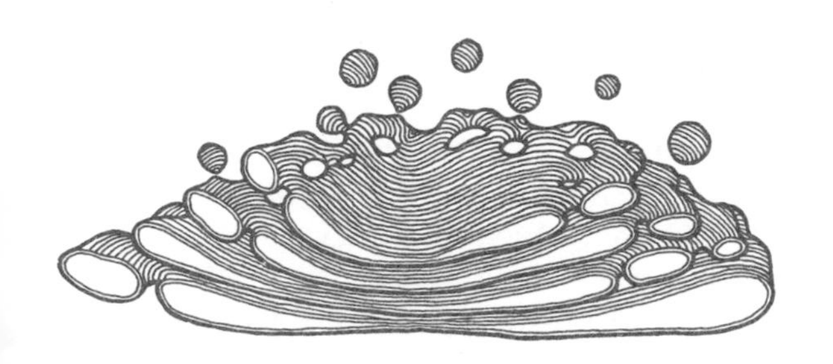

Opposite: an electron micrograph of a section through an organelle known as the Golgi apparatus – magnified 60,000 times. Left: a three-dimensional drawing showing how the Golgi apparatus packs enzymes into membranous capsules before releasing them. Above: a mucus-secreting intestinal cell in which all cell functions are subordinated to that of the Golgi apparatus

the cell's outer membrane and the endoplasmic reticulum – are all very similar. The membranes of the mitochondria and chloroplasts are also of the same general type as is found in the rest of the cell. For this reason there is considered to be one basic type of cell membrane, called the unit membrane. Under the electron microscope these membranes appear as a double line of dense material with a space of transparent material in between. The dense material on the outside is thought to be protein and the light material in the centre of the sandwich to be fat. Membranes from different parts of the cell do, however, vary in thickness from about 0·007–0·01 micron.

About one third of the membrane by weight is fatty material and most of the rest is protein. There is no general agreement at present about the way in which the protein and fat are organized in the membrane. Fat-soluble substances readily penetrate the cell membrane, but so do water-soluble substances. There are also some quite simple molecules which cannot penetrate the cell. Some substances are taken into the cell actively and reach much higher concentrations within it than in the surrounding liquid. This process, called active transport, is performed by enzymes called permeases and is an energy-requiring process. There may be holes in the membrane which allow some substances to pass through and not others. Alternatively, the membrane has been considered to be a mosaic of protein and fat with areas where fatty substances can dissolve through the membrane and other areas where water-soluble substances can dissolve through it. Both the fat and the protein are necessary for the integrity of the cell membrane, since if agents which dissolve fats or denature proteins are applied to the membrane, it is destroyed.

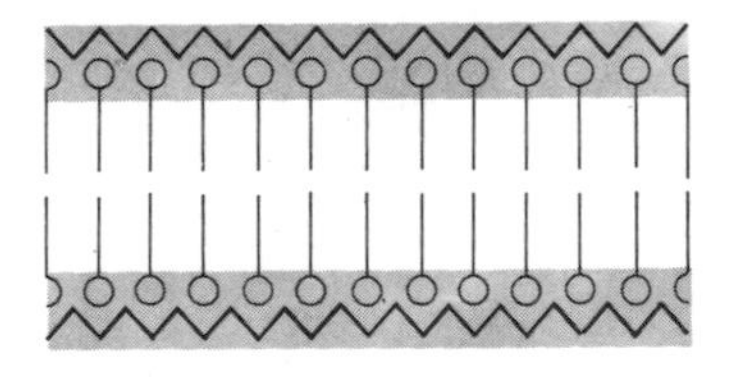

All cell membranes appear to have the same basic structure – two one-molecule thick layers of proteins enclosing a two-molecule-thick layer of fat. The shaded areas in the diagram indicate how this, the unit membrane, acquires its distinctive 'double' appearance in electron microscope pictures

Lysosomes – the 'suicide bags'

In 1949, Christian de Duve, while studying enzymes extracted from rat liver in his laboratory in Louvain, Belgium, was unable to explain why the quantity of one enzyme varied in apparently identical experi-

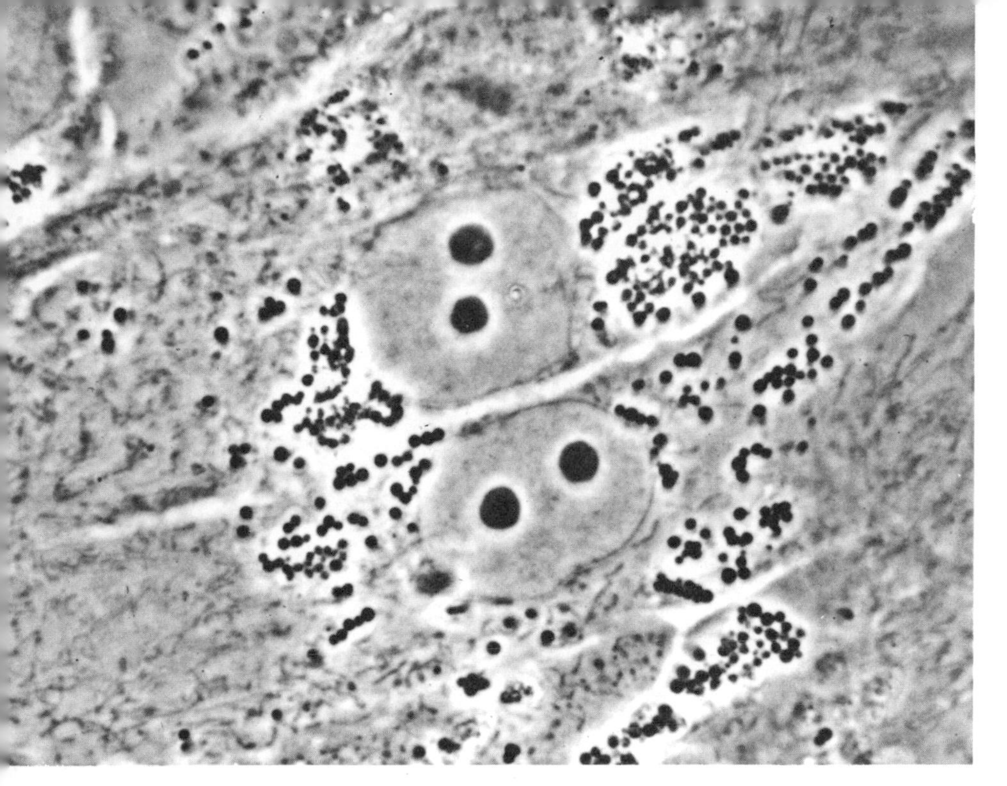

Lysosomes – several hundred of them – show up clearly in this specially stained micrograph of kidney cells

ments. Then he noticed that when he homogenized the rat liver cells gently he obtained much less of this enzyme than when he homogenized the same cells vigorously. This was the first clue to the existence of a new cellular particle – the lysosome. The particles were given this name because they contain highly active enzymes able to destroy, or lyse, the cell.

Lysosomes are little membranous sacs which contain a battery of enzymes. These sacs are easily broken when the cell is homogenized. They are from 0·25–0·8 micron in diameter and contain enzymes capable of digesting proteins, RNA, DNA and carbohydrates. The lysosomes may be thought of as apparatus for handling and packaging dangerous and corrosive chemicals, for so long as the lysosome membrane is intact the enzymes cannot damage the cell. However, when the biochemist breaks open cells for experimental purposes by grinding them, the lysosomes may also be broken down and so release enzymes which damage other parts of the cell unless great care is taken.

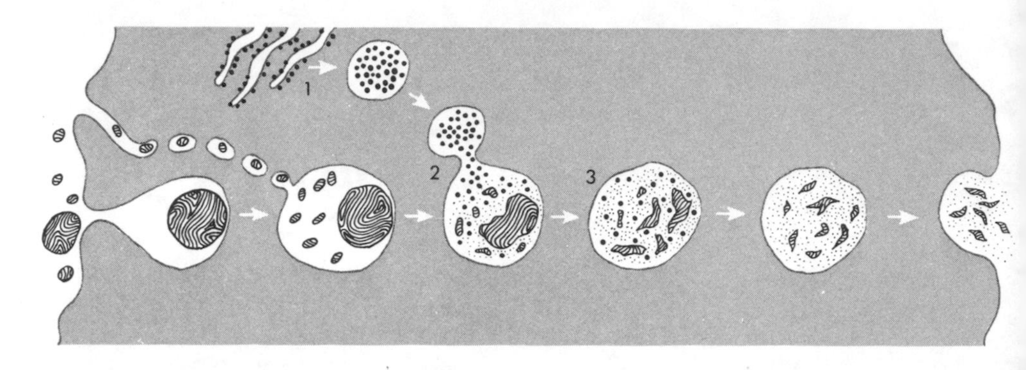

Cell digestion. Ribosomes (1) associated with the endoplasmic reticulum form a lysosome, which fuses (2) with a food vacuole. In the digestive vacuole so formed (3) lysosomal enzymes break down the trapped food particles and digestive products are absorbed through the surrounding membrane, until finally the vacuole's residual contents are expelled

One of the functions of the lysosomes is to digest pieces of food taken in by the cell. This is done in special food vacuoles. The white cells of the blood (phagocytes), for example, have an important function as scavengers. They engulf pieces of dead tissue, or perhaps whole bacteria, and destroy them with the assistance of lysosomes, which release their digestive enzymes into the vacuole, or space, in the cytoplasm, surrounding the engulfed material. The membrane round the vacuole prevents the digestive enzymes from damaging the cell itself. Protozoa such as *Amoeba* and *Paramecium* also digest food by the liberation of lysosomes into vacuoles containing food particles.

Worn out parts of the cell, too, are digested in similar vacuoles, and the molecules of which these worn out parts are made are put back into circulation once more as food. In an old cell the lysosomes may dissolve and result in the cell's complete breakdown – for which reason they are sometimes called 'suicide bags'. The lysosomes play an important part in the death of cells in the course of embryonic development. When a tadpole develops into a frog its tail is 'reabsorbed' – a result of the action of lysosomes in the tail cells which literally digest away the tail from within. Cell death brought about by lysosomes is now known to be a normal part of the processes by which tissues and organs are remodelled during embryonic development.

The lysosomes may play an important part in some disease processes as well as in disease resistance. Silicosis, a disease common among miners, caused by the inhalation of silica dust, may result from an accumulation of these silica particles in the lysosomes in lung cells, and so cause corrosive enzymes to leak out of the lysosomes. Excessive quantities of other substances, such as vitamin A, are also known to cause the lysosome membranes to become unstable. Other substances such as cortisone and hydrocortisone stabilize the lysosome membrane, and this may account for the well known anti-inflammatory properties of these substances.

Mitochondria and chloroplasts

The mitochondria (plural of mitochondrion) and the chloroplasts might be described as the batteries or powerhouses of the cell. There may be anything from one to more than a thousand mitochondria in a single cell. They vary in diameter from 0·5–1·0 micron and may be up to about 7 microns long. In the mitochondria pyruvic acid (derived from sugar) is burned, so releasing energy for use by the cell. The energy is used to synthesize molecules of adenosine triphosphate (ATP) which diffuse out of the mitochondria to all parts of the cell. ATP, often called high-energy phosphate, releases its energy in serving the needs of the cell and is destroyed in the process. All biochemical reactions are reversible, but a reaction may be, and often is, much slower in one direction than the other. ATP speeds up reactions by supplying energy so that they occur in the direction and at a rate which is useful to the cell.

The mitochondrion is a highly organized double-walled sac. The outer membranous wall is smooth, but the inner membrane is full of convolutions and deformations which stretch right across the interior of the organelle. These deformations, called cristae, stick out from all sides of the inner membrane. Although under the electron microscope both these membranes appear similar to other cell membranes, in fact they differ a great deal biochemically. They are made up of a mixture of protein and fatty material held together by what are called hydrophobic bonds – chemical bonds

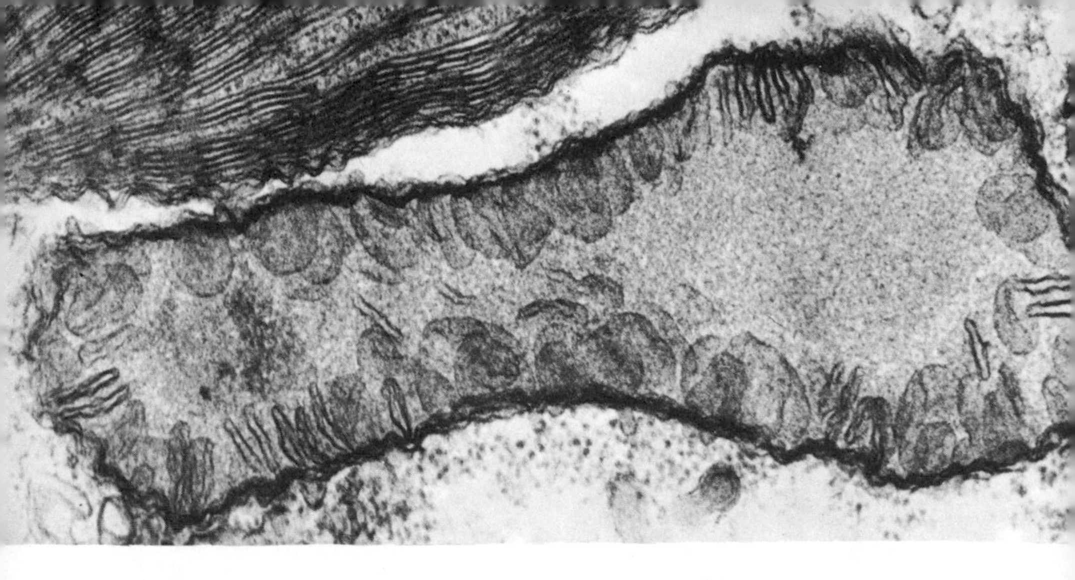

The cell's chief supplier of chemical energy. Section through a mitochondrion magnified 50,000 times. The inward-jutting structures are the cristae, which are concerned with the synthesis of ATP from sugars

which are not water-soluble. The protein part of the membrane consists of some proteins which have a structural function and others which have an enzymic function.

The synthesis of ATP by the mitochondrion is brought about by the conjunction of two separate biochemical pathways, each involving many enzymes. These two separate biochemical pathways are called the Krebs citric-acid cycle and the electron-transfer chain. The citric-acid cycle is a collection of about ten enzymes which work together to remove energy from pyruvic acid molecules in a series of easy stages. This might be thought of as a process of slow and controlled burning in which the heat energy is trapped in ATP molecules. Actually the energy is liberated in the form of energy-carrying hydrogen and electrons which are then passed along the electron-transfer chain so that ultimately hydrogen combines with oxygen to form water and the electrons give up their energy to assist in the synthesis of ATP. This type of energy release is called aerobic energy release: that is, it uses oxygen from the air. Anaerobic (without air) energy release is effected by some organisms, such as yeast, by means of alcoholic fermentation, which is carried out by enzymes in the cytoplasm, outside the mitochondria.

The citric-acid cycle enzymes are believed to occur

in solution in the centre of the mitochondrion between the cristae. The enzyme molecules of the electron-transfer chain are located in the membranes of the cristae themselves. High-power electron microscope pictures of the cristae show structures shaped like a knob on the end of a stick studded all over the inner surface of the cristae. These are molecules of the enzyme adenosine triphosphatase, which catalyses the final synthesis of ATP from AMP (adenosine monophosphate) and inorganic phosphate.

Until recently there was a great deal of dispute about whether mitochondria were formed by division of other mitochondria, whether they derived from other structures called microbodies, or whether they budded off from the nuclear membrane. Now, however, DNA has been discovered in mitochondria, and it has become clear that mitochondria normally arise by the division of other mitochondria. (Yet, it was known for many years before this discovery that mitochondria had hereditary properties.)

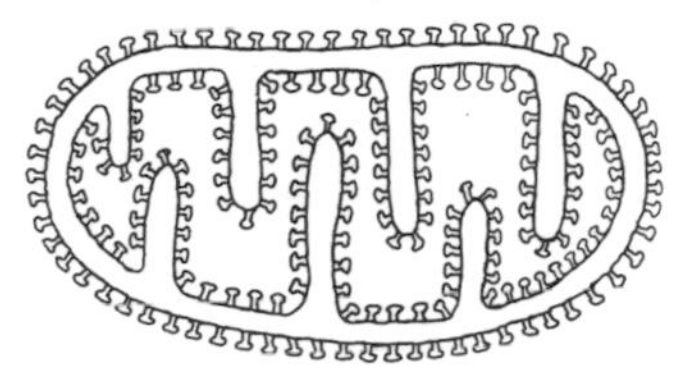

Diagrammatic section through a mitochondrion showing how the cristae arise as infoldings of the inner membrane

Mitochondria are absent in certain yeast cells which give rise to strains of yeast that grow very slowly and form only very small colonies on culture plates. These strains are called *petite colonie*, or just petite. Yeast is exceptional among cells with a well-defined nucleus in being able to grow completely anaerobically and in not being entircly dependent on the efficient functioning of its mitochondria, as many other cells are. The petite character is inherited in petite-cell lines indefinitely, without any reversions to the wild-type yeast ever occurring. However, if petite cells are mated to normal cells all the progeny are normal. The petite character is lost in the process of mating, as if normal mitochondria were restored to the petite cells by the mating process. This type of inheritance, called cytoplasmic inheritance, is quite different from the processes of inheritance involving nuclear genes which, in similar circumstances, would be expected to give two types of cells – petites and normals – in equal numbers in the progeny. In fact, a mutation of a nuclear gene occasionally turns up which looks the same as the cytoplasmically inherited petite, but differs in showing nuclear inheritance. Other organisms which have distinct male and female reproductive cells always inherit the cytoplasm from the egg cell.

For a long time the hereditary properties of mitochondria were considered to be a curious anomaly by some, and were a source of embarrassment to others who believed that the nucleus was the repository of all genetic information and all the DNA of the cell. However, sound evidence began to accumulate in the mid-1960s for the existence of DNA in the cytoplasm. The quantity of DNA in the mitochondria is in fact quite small – probably just about enough to code for three or four average-sized proteins. It is the amount of DNA which might be found in a rather small virus. The mitochondrial DNA is circular, like the DNA of many viruses and bacteria. It is not known for certain which proteins of the mitochondria are coded for by the mitochondrial DNA. However, since there are very many more than five different proteins in the mitochondrion, the majority of these mitochondrial proteins must be coded for in the nucleus and synthesized in the cytoplasm, before migrating to the mitochondrion.

The mitochondrion appears to have a system of protein synthesis which is different from that of the cell – a system sensitive to inhibition by certain antibiotic drugs which do not affect the normal protein synthesis in the endoplasmic reticulum. These same antibiotics also inhibit protein synthesis in bacteria; and in fact there are other features of the protein synthetic machinery of the mitochondrion which bear a close resemblance to bacterial protein synthesis. The DNA of the mitochondrion also resembles bacterial DNA more closely than it does the nuclear DNA of animal and plant cells, not only in its circularity, but also because it lacks a coating of histone protein.

The resemblance of mitochondria to bacteria has suggested to many biologists that mitochondria may have evolved from intracellular parasitic bacteria. Many protozoa contain in their cytoplasm living bacteria and algae which were originally parasites, but which later evolved a useful relationship with the host cell. Protozoa containing photosynthetic algae, for example, can be kept alive in the presence of light without any external food supply, since nutrient molecules synthesized by the algae are released into the cytoplasm of the protozoa. Indeed, some of the intracellular symbionts of protozoa have become so specialized that they are no longer able to live outside the host cell. It seems quite possible that mitochondria may have evolved in a similar way – though we cannot be certain, for it is also possible that mitochondria may have evolved from some part of the cell nucleus.

Chloroplasts are similar to mitochondria in containing DNA and in having their own characteristic protein synthetic system. The origin of chloroplasts is similarly a subject of speculation. It is possible that mitochondria and chloroplasts have evolved from the same primitive particle – a particle which might, in the early stages of its evolution, have performed the energy-converting functions of both mitochondrion and chloroplast.

Chloroplasts, which are found only in plant cells, are the organelles which contain chlorophyll, perform photosynthetic activities and, of course, give plants their green colour. Photosynthesis is the process by which the sun's energy is absorbed and trapped in chlorophyll pigment molecules. This energy is used to synthesize ATP and ultimately sugars which can be stored by the cell. The energy stored in the sugars can then be used by the cell as it is needed. This energy storage is in principle the reverse process to the energy breakdown which occurs in the mitochondria, but it is performed by an entirely different set of enzymes.

Chloroplasts sometimes show hereditary differences and, as with mitochondria, these are always inherited through the cytoplasm of the mother. The pollen or other male fertilizing cell contains only a little cytoplasm and this does not usually seem to contribute to the fertilized cell.

The chloroplast is usually 4-6 microns in diameter

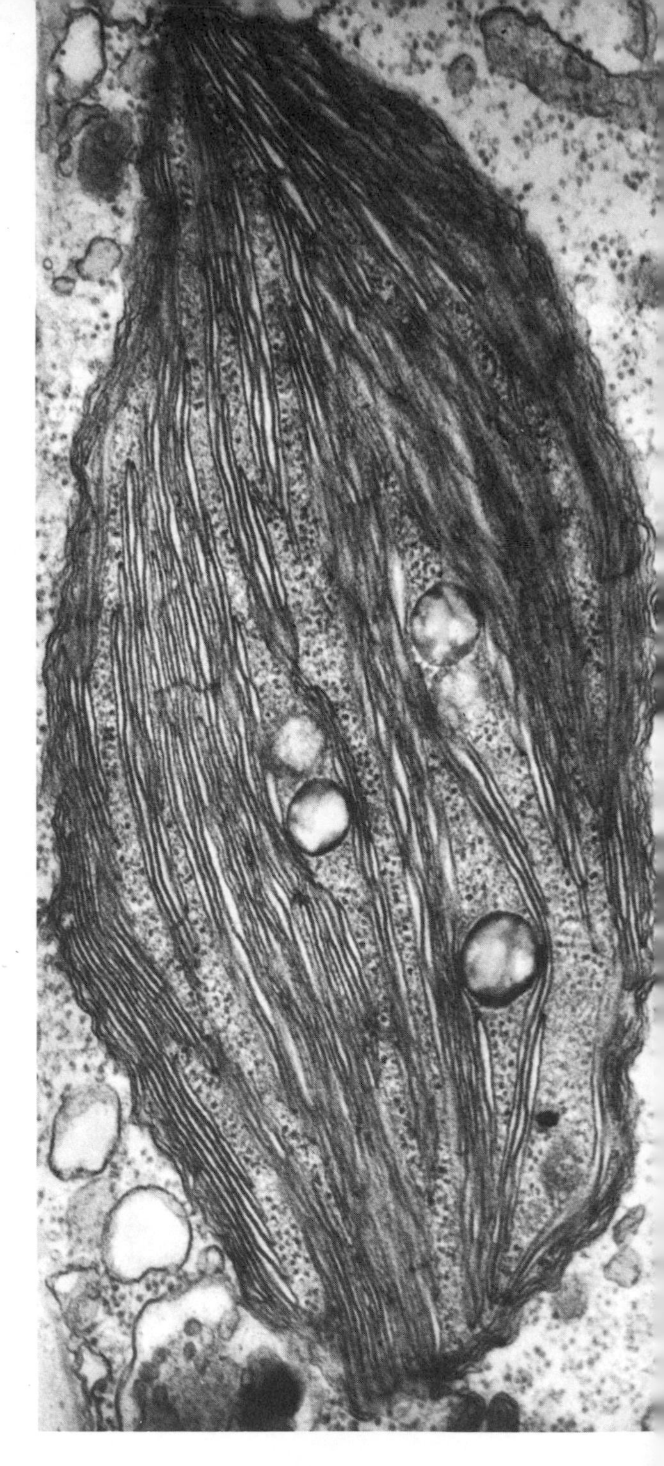

The site of photosynthesis. Section through a chloroplast, enlarged 50,000 times. The active centres of photosynthesis – the grana – are oriented longitudinally. The prominent round objects are merely globules of fat

and can be spherical, ovoidal or even disc-shaped. There can be one chloroplast per cell, as in some algae, or hundreds, as in some higher plants. The chloroplast consists of about 56 per cent protein, 32 per cent fat and about 8 per cent chlorophyll. In electron microscope preparations the chloroplast can be seen to be made up of large parallel strands, or lamellae. Sometimes a series of extra 'plates', called grana, is present between the lamellae. On the surface of the grana there can be seen small bodies called quantosomes, which each contain 250 molecules of chlorophyll.

The structure of the chloroplast changes if it is kept in the dark: the chlorophyll and the lamellae disappear, but reappear if the chloroplast is again placed in the light. Chloroplasts can be formed from the division of pre-existing chloroplasts or from submicroscopic bodies called proplastids, which grow to about one micron in size and then begin to form lamellae from their outer membrane. There are several other types of plastid present in the plant cell besides chloroplasts. There are leucoplasts, in which starch grains develop as a food store, and chromoplasts, which contain pigments other than chlorophyll.

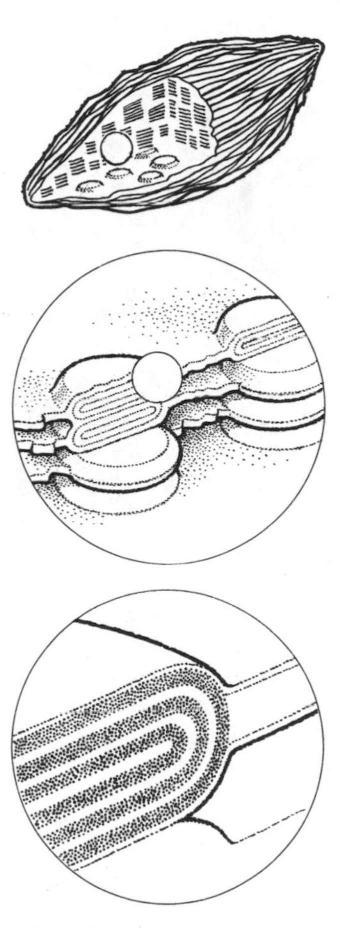

Chloroplast structure. Top: a cutaway drawing of the whole organelle, locating the grana. Middle: the grana turn out to be a battery of quantosomes, usually compared in appearance to a stack of coins. Bottom: a section through a quantosome, with the key layers of chlorophyll molecules indicated

Other organelles

Many cells, but probably not the cells of higher plants, contain a body near the nucleus called the centrosome. The centrosome actually consists of two separate bodies about a fifth of a micron in diameter called, with their associated cytoplasm, the centrioles. When the cell divides the centrioles move to opposite ends of the cell and appear to play an important part in the formation of the spindle on which the chromosomes are arranged at cell division. Electron microscope pictures show that usually each centriole is a paired structure, rather like two cylinders at right angles to each other. In fact it looks rather as though new centrioles form by a process of growth at right angles to the parental centriole. In cross-section the centriole can be seen to consist of nine filaments placed at the edge of a cylinder, each of these filaments consisting of three associated fibres.

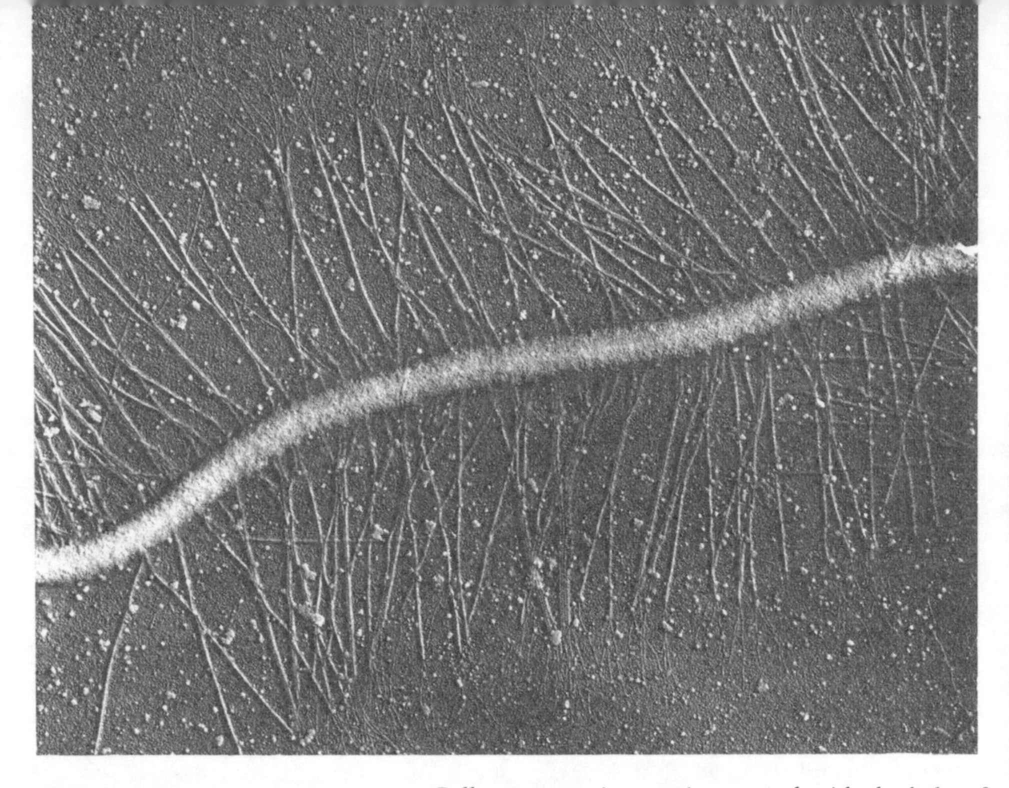

By lashing a flagellum or by sweeping with cilia many cells are able to move through, or to generate a current in, a liquid environment. Above: a length of the flagellum of the zoospore Bumilleria, magnified 16,000 times. Its two rows of lateral hairs enhance its efficiency. Cilia are smaller than flagella and usually occur in groups, acting in concert. Opposite: a section through a number of adjacent cilia. Here the magnification is 96,000 times

Cells can sometimes swim around with the help of little thread-like organs called cilia or flagella. The tail of a sperm is an example of a flagellum; cilia are distinguished from flagella by being shorter and occurring in large numbers on one cell. Flagella are present in relatively small numbers, there often being only one flagellum present. Many cells have neither cilia nor flagella. Both cilia and flagella grow out from a basal granule closely resembling a centriole. The respective structures of cilia and flagella as seen under an electron microscope are sufficiently similar to the structure of centrioles to suggest that cilia and flagella may have evolved from them. Almost all cilia and flagella when seen in cross-section under an electron microscope consist of nine double fibres arranged in a circle with one double fibre in the centre. There are, however, some departures from this basic pattern which show that the pattern is flexible and so might well have originated as a structure similar to the centriole. The basal body, or kinetosome, of cilia and flagella contains

DNA and is inherited as a separate organelle, like a mitochondrion or a chloroplast. Cells can be permanently deprived of their kinetosomes and associated flagella by drug treatments – which shows that the nucleus of the cell does not contain all the information for the synthesis of the flagella. The DNA of the kinetosome must also contain indispensable information.

Boxes within boxes

The cell could be thought of as a nest of chinese boxes – each box representing a different level of organization, so that if one box is removed there is always another inside it. The first box represents what we can see with the light microscope – the nucleus and the larger organelles in the cytoplasm. The second box represents what is revealed by the electron microscope – there is much more detail and some of the larger molecules of the cell can just be distinguished. Biochemical experiments tell us how these molecules behave in the living cell, and X-ray studies can reveal the actual structure of the molecules themselves and even pinpoint the positions of the atoms of which these molecules are made.

When a child explores a nest of chinese boxes he eventually comes to the last box and finds it empty. This invariably disappoints him because his illusion of an infinite series has been destroyed. Scientists exploring the cell have yet to reach this point. The second box has only recently been opened and it is as though the boxes beneath were transparent. The general pattern of the underlying molecules and cell parts can just be discerned, but many of the details still elude us. Only when all the boxes have been separated and examined will we be in a position to understand exactly how the living cell works and how it has evolved from a simple primitive cell and, ultimately, from inorganic matter. The history of this evolution is almost certainly still recorded in the detailed structure of the many parts of the cell, but particularly in the structure of the protein molecules themselves.

HOW CELLS REPRODUCE 4

Could a robot – or any machine – capable of reproducing itself like a living thing be constructed by man? For a long time it was thought that such a machine would need an impossibly large number of components. In 1958, however, Professor L. S. Penrose of London University succeeded in building just such a machine – although reproducing itself was about all it could do.

Penrose's machine consists of ingeniously shaped wooden blocks which can hook together as pairs in a special way. When joined pairs of these blocks are shaken together in a tray with separate unit blocks they hook up with the units and then split apart again to form more joined pairs. In this way a single pair of joined blocks can multiply to form an indefinite number of other pairs from individual units. The tray must of course be shaken to provide energy for the joining process to occur – but this could be compared with the light energy which must be supplied to plants for them to grow and multiply. It is not in any way inaccurate to say that these paired blocks are self-reproducing machines – primitive robots.

Whether we look at the DNA molecule, at the chromosomes, or at the cell itself, biological reproductive processes are not very different in principle from those involving Professor Penrose's blocks. Cells reproduce themselves by an exact copying and splitting process which results in two *identical* daughter cells – providing there are no accidents. These daughter cells

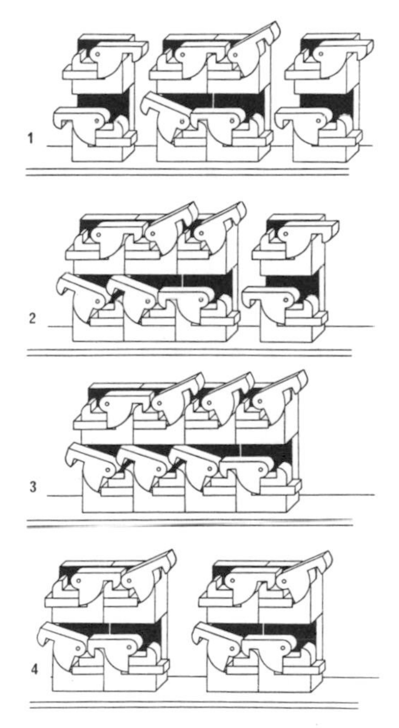

Biological reproduction is not creation out of nothing, but simply an ordering or re-ordering of already existing units. The Penrose model (above) illustrates the principle. A two-unit entity (1) links with a third unit (2). Linking with a fourth unit, however, causes the original pair to separate (3), leaving two identical two-unit entities (4)

have identical chromosomes with the same genetic message coded in their DNA and a sufficient share of cytoplasm and cell organelles to give them a good start in life. This process of simple multiplication in numbers is termed asexual reproduction.

Sexual reproduction is much more complicated and for it to occur two different parent cells must first fuse. The chromosomes and cytoplasm mix together to give the new cell a mixture of the parental characteristics. In fact the cytoplasm may all be contributed by one cell – the egg – and the sperm may in effect only contribute its chromosomes. The chromosomes of egg and sperm line up together and re-arrange themselves according to a set plan, so that each daughter cell receives a complete set of chromosomes, half of which comes from each parent.

In this way the sexual process ensures a regular shuffling of parental characteristics – or, more precisely, of the genes on the chromosomes. The sexual process also ensures that the species contains a variety of different individuals which share common characteristics. Some of these individuals will be fitter than others to withstand the rigours of the environment and will survive to perpetuate the species. In this way sex provides an evolutionary advantage to those species which reproduce by means of it – perhaps that is why sex is found in the whole range of living things from bacteria to higher animals and plants. Nevertheless, there are some conservative, or should I say unadventurous, organisms which do not reproduce sexually. These organisms can only evolve very slowly, by mutation. Common examples of such organisms are the amoeba and the euglena. Many algae and most fungi also reproduce asexually.

How are the processes of sexual and asexual reproduction achieved without losing any of the essential genetic material? The cell goes through an elaborate performance involving the pairing of like chromosomes, after which the chromosome pairs are shuffled – in the case of 'sexually dividing' cells – and separated like a pack of cards.

Opposite: cell replication by mitotic, or asexual division. As the time of division approaches, the nuclear membrane and nucleolus become less visible, while the chromosomes condense and line up in an equatorial plane. The chromosomes then divide (DNA replication occurs before cell division begins) and identical sets of chromosomes migrate to the nearest pole; whereupon the cell divides into two daughter cells, both identical to the parent cell from which they came

Asexual cell division – mitosis

The cell, when it is not dividing, has a nucleus in which very little, if any, structure may be seen. No chromosomes are visible in it, although pieces of a thread-like network can sometimes be seen attached to the nucleoli and nuclear membrane. The early microscopists used to call this stage the 'resting' cell, because no chromosomes could be seen moving in the nucleus. It is still often called the 'resting' cell at this stage, even though the cell is in fact very busy growing. The DNA of the chromosomes is unravelled in the 'resting' cell and is occupied with the task of providing messenger molecules for protein synthesis.

When the cell starts to divide, the nucleoli disappear and the chromosomes begin to condense and become visible. In asexually-dividing cells the chromosomes can be seen to be double, although the two parts – the chromatids – are linked together at a constriction called the centromere. There may be anything from one to a hundred chromosomes present in a single cell. How the chromosomes and the rest of the cell behave during asexual cell division (called mitosis) is illustrated diagrammatically.

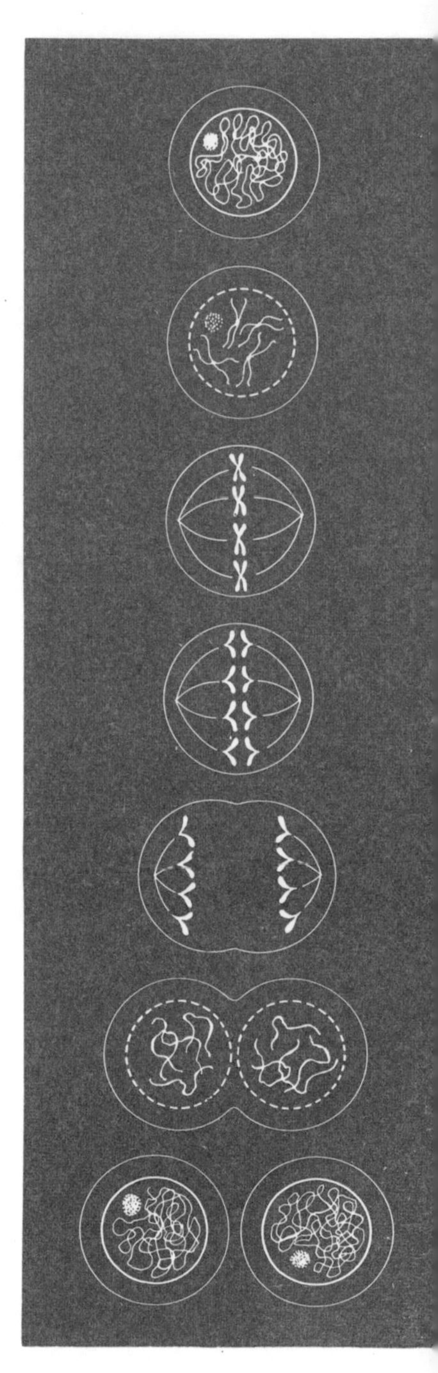

Sexual cell division – meiosis

Every normal cell in the body of an animal or plant contains the same number of chromosomes – usually an even number. This is because each chromosome has its double, or homologue – one homologue coming from the father and one from the mother, so giving an even number of chromosomes. When the reproductive cells – the egg and sperm – are formed, the number of chromosomes present in each must be reduced to half the number ordinarily present in body cells; otherwise with each generation the number of chromosomes per cell would double – a situation which obviously could not continue or the cell would become indefinitely large. In the sexual division of the cell (called meiosis), therefore, the chromosomes are juggled about so that each of the daughter cells receives only one homologue of each chromosome,

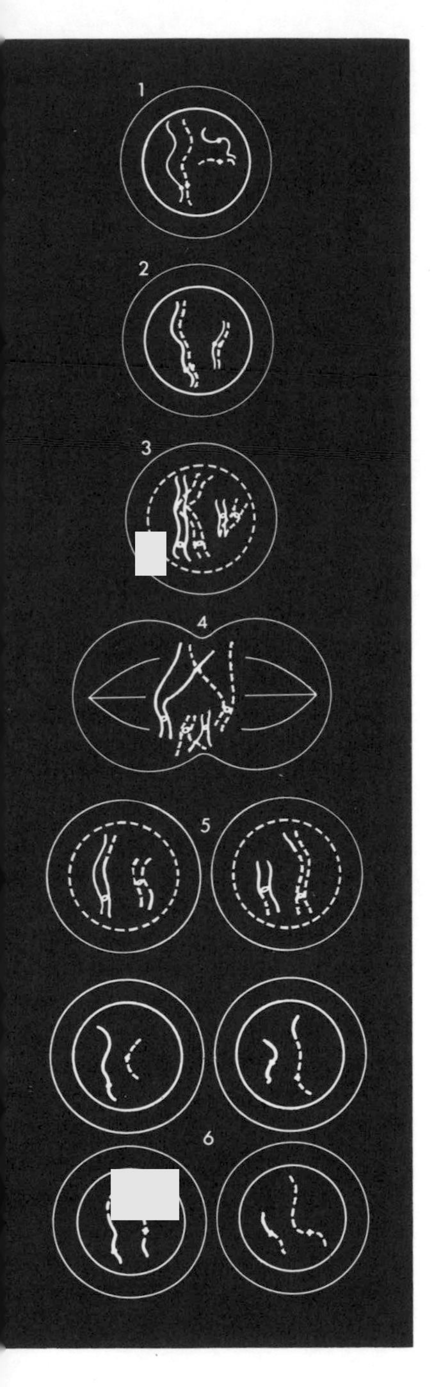

thereby halving the total number of the chromosomes that would otherwise be present. If the body cells of the adult organism each contain $2n$ chromosomes (called the diploid number), then the egg and sperm cells will contain n chromosomes (called the haploid number). The actual number n varies with different species.

Sexual cell division is also sometimes called reduction division, because of the reduction in chromosome numbers. The cell achieves this reduction by dividing twice while the chromosomes only divide once. How this actually happens is shown diagrammatically: one cell with $2n$ chromosomes has divided to produce four cells with n reshuffled and recombined chromosomes. Each cell has all the genes it needs to synthesize all the cell proteins, but each cell is genetically different from the others, carrying its own unique combination of the genes from the parent cell. It is a remarkably neat and precise way of exactly halving the cell DNA and at the same time reshuffling the genes. Such haploid sex cells are capable in some organisms of living independently – in others they exist only as eggs or sperms.

Fertilization

In primitive organisms there may be no distinct differentiation into male and female germ cells. Many primitive organisms – for example, protozoa, fungi and bacteria – have what is called a mating-type system which serves the same sort of purpose as sex. If an organism is of mating-type A then it can only mate with an organism of mating-type B. A cannot mate with A, nor B with B. This ensures that brother and sister cells do not breed together. Outbreeding – the breeding of unrelated organisms – is usually advantageous since it allows new hereditary combinations to arise more easily. In many organisms the existence of two sexes ensures that outbreeding will occur. However, there are many bisexual flowering plants which bear the male and female parts on the same individual, so that self-fertilization might easily occur. These

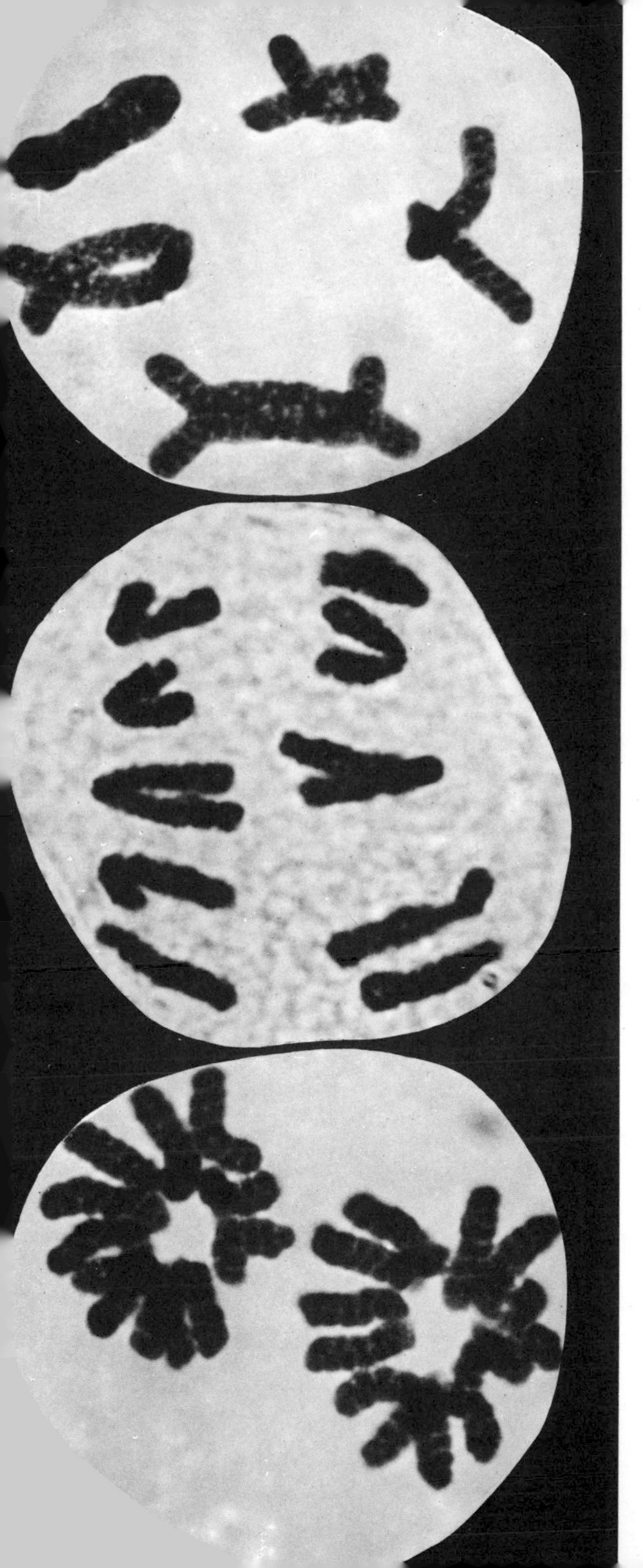

Opposite: the meiotic cell divisions that occur during the formation of eggs and sperms. The chromosomes appear as distinct entities (1) which pair off (2). The chromosomes of each pair (one shown in straight, the other in dotted lines) are similar but not identical. Each chromosome separates into two chromatids (3) and 'crossing-over' of genetic material occurs (4); two non-identical daughter cells result (5). Both cells subsequently undergo a further division which has the effect of halving the normal number of chromosomes (6) producing four distinct and unique cells. The normal number is restored if and when one of these cells fuses with another reproductive cell from the opposite sex at fertilization

Left: stages during the first meiotic division, observed in a lily cell. Top to bottom: chromosomes dividing; migrating towards opposite poles after division; and clustering at opposite poles to form two separate nuclei

The human sex cells. Right: an electron micrograph of a male sperm reveals a head, dense with chromosomes, with flagellum attached. The organelles round the upper part of the flagellum are mitochondria. Below: a fertilized female egg cell. The egg and sperm nuclei, which have not yet fused, can be seen distinctly

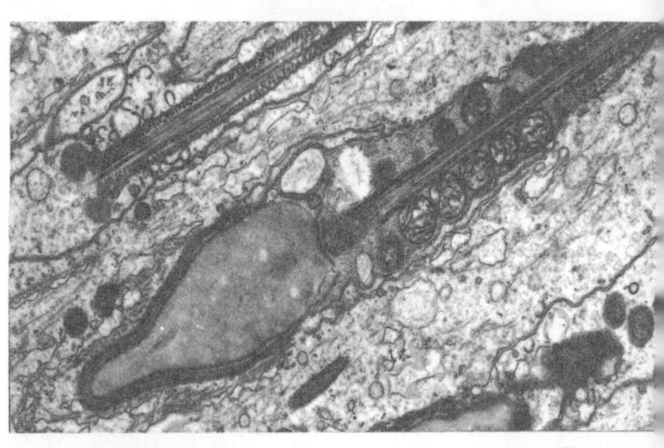

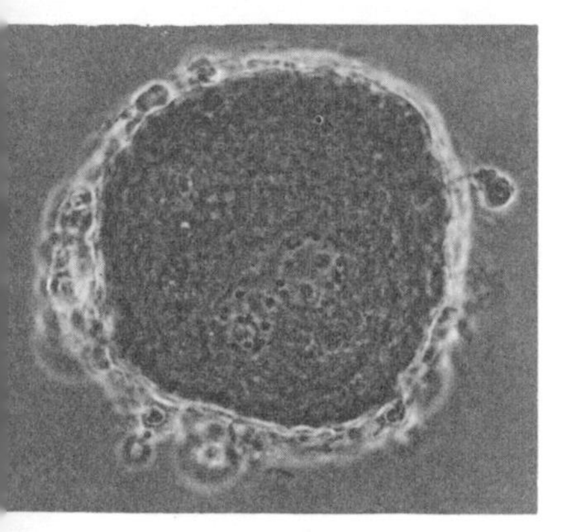

plants often have a mating-type system, as well as a bisexual constitution, to ensure the occurrence of outbreeding.

Sex in its stricter biological sense is usually used to refer to the situation in which a difference in size exists between the germ cells so that the smaller usually motile germ cells are called sperm and the larger stationary germ cells eggs. Sex in this sense is found in quite primitive organisms, such as *Volvox*, showing that sex may have arisen early in the evolution of living things. Both egg and sperm cells are formed by the process of reduction division just described. However, in the case of the egg three of the four cells produced – the polar bodies – are very small and are discarded, leaving only a single egg in possession of nearly all the cytoplasm. The process of discarding the polar bodies, to leave an egg with almost four times as much cytoplasm as it would otherwise have, makes for a larger egg – so giving the offspring a better start in life.

In the case of the sperm, all four cells produced by the reduction division develop tails and become mature sperms. The female therefore supplies a large immobile egg full of nutritious yolk and a set of chromosomes, while the male supplies, in the sperm, a streamlined delivery vehicle containing a second set of chromosomes and little else. Sperms are usually produced in much larger numbers than eggs because of the enormous wastage of sperms prior to fertilization.

The fertilization of the egg is the reverse of the reduction division. The two haploid chromosome sets (n) from the sperm and egg join together to give the diploid number of chromosomes ($2n$). As soon as one sperm has entered the egg there is a change in the egg membrane which prevents penetration by other sperms. How the fertilized egg develops into an embryo will be described in chapter 8.

Multiplication and growth

A bacterium, such as the colon bacillus, growing at its fastest, may double in size and weight every twenty minutes. The cells of other organisms may take very much longer to do this – *Amoeba proteus*, for example, divides only once every twenty-four hours. Certain human cancer cells – called hela cells after a woman named Helen, from whom samples were first taken – can divide every ten hours when grown in the test tube. The actual process of cell division only occupies a fraction of the time that elapses between divisions, perhaps ten per cent of it, the rest being taken up by growth and preparation for the next division. After the cell has divided there is a further period of growth and reorganization, and then a period during which the DNA doubles within the cell. This is then followed by yet another period of growth before division finally occurs.

Microbial cells will grow indefinitely if transferred to a new medium from time to time – so they might be said to be immortal. A cell divides into two, and each of these daughter cells divides again to give 4; these divide to give 8, 16, 32, 64 and so on. This increase in the number of cells is described by a simple equation $N = 2^n$, where N is the number of individual cells and n is the number of cell divisions. As n increases, so N also increases, doubling with each generation. So long as there is food available most micro-organisms will multiply indefinitely, doubling at a constant rate. Thus, after 10 cell divisions occurring in only 200 minutes, a single colon bacillus can multiply into 1,024 cells; and after 20 divisions (taking less than 7 hours) the same bacillus will have multiplied into more than a million cells.

Cells in the tissues of multicellular organisms cannot normally multiply indefinitely in the body of the organism because the size of all tissues is controlled by means of hormones. Cancer cells, however, are exceptional in that they manage to evade the control systems of the body. Cells from multicellular organisms can be grown in tissue culture indefinitely, provided the culture medium is regularly changed. In one experiment, chick heart cells were cultured continuously through two world wars – from 1913 to 1946.

It is not known exactly why cells divide when they reach a certain size. If an amoeba cell is shaken while it is dividing, it divides unevenly to give one large and one small cell. However, both cells subsequently grow to the same critical size before they divide again – the smaller cell of course taking longer to do so. If a piece of cytoplasm is chopped off an amoeba just before it divides, it continues to divide regardless, producing daughter cells smaller than normal; but if a piece of cytoplasm is chopped off an amoeba while it is still growing, it continues to grow to the normal size before it divides. These experiments show that the actual quantity of cytoplasm, or the surface-to-volume ratio of the cell, is not the most important factor in triggering cell division. It is not known whether there is a single trigger – some cellular hormone, for instance – to start cell division. It seems more likely that many preparatory processes must run their course before the cell will normally divide. Among these are probably the doubling of the DNA and the doubling of cell size.

Heredity and genes

The genes have been mentioned as instructions – encoded in the DNA of the chromosomes – which specify the nature of the proteins which are synthesized in the cytoplasm. This is how genes are understood now; originally they were conceived as rather abstract entities, on the strength of experiments involv-

Opposite: members of a species of Euglena, all apparently in the final stages of mitotic fission. Exactly what mechanism causes cells to divide when they do is not yet clearly understood

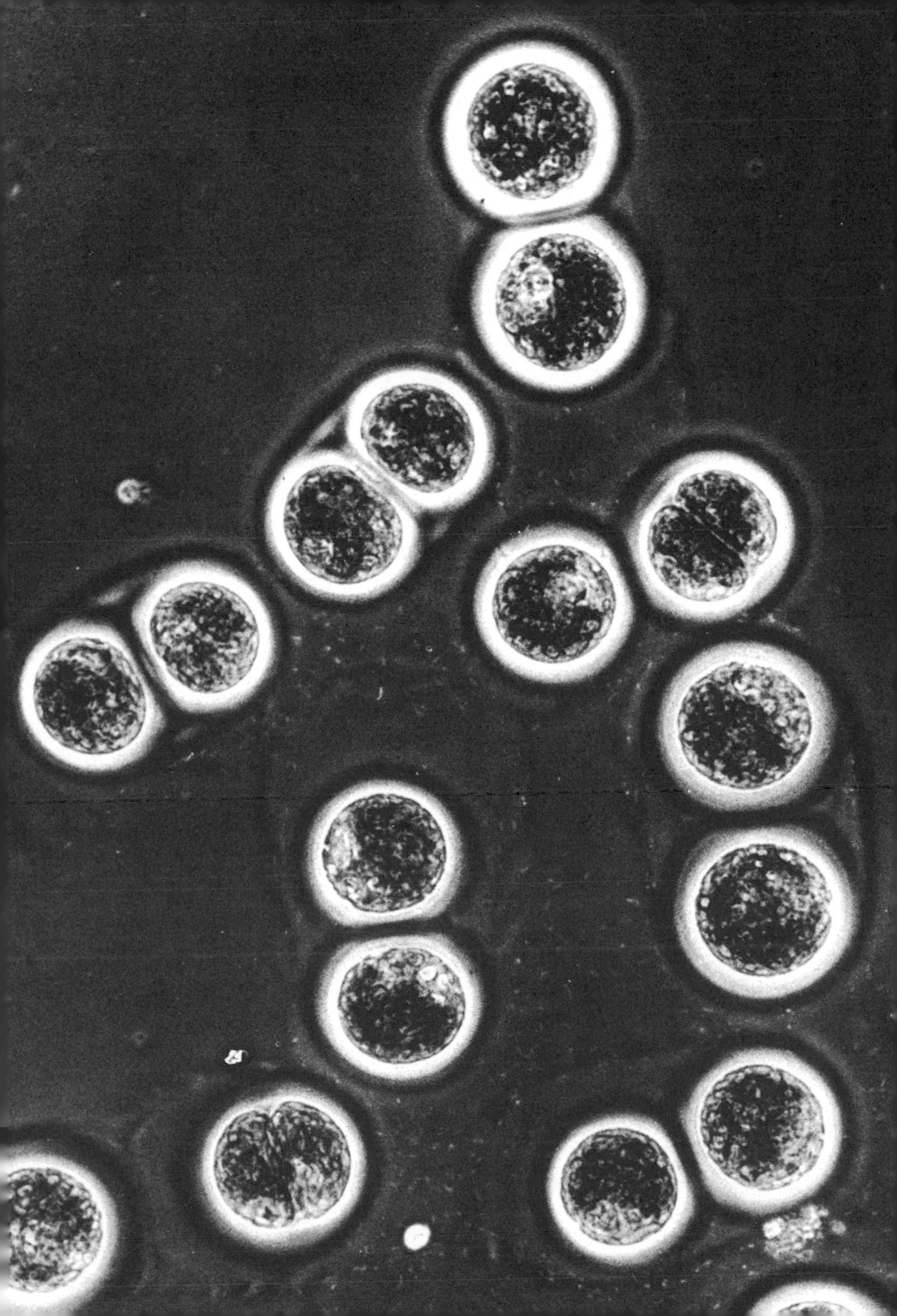

ing crosses between different strains of peas. Gregor Mendel, a monk living in Brno in what is now Czechoslovakia, was the first man to understand the laws of inheritance. His experiments are now recognized as being among the most brilliant and original contributions to biology, yet his contemporaries failed to appreciate their significance. Mendel's work was published in 1865, but its importance was not recognized until 1908, when it was rediscovered by three scientists simultaneously.

Mendel deliberately chose to work with peas because they could easily be fertilized artificially, and many distinct cultivated varieties were available. He chose to investigate those aspects of the peas which he found to show clear-cut differences in different varieties. He studied the colour of the seed and of the unripe pods, the shape of the pods and the seeds, and the height of the plants when there was a clear difference between dwarf and giant varieties. He was careful not to study small inherited differences in height. This was just as well, since such differences do not show clear-cut inheritance due to a single gene, but have been found to be the result of the action of large numbers of genes with small individual effects. Mendel noted that there was a great deal of variation in height within the short and tall varieties that he studied, but this variation did not overlap. Short plants were 9–12 inches tall and tall plants 6–7 feet.

The Moravian monk Gregor Mendel (1822–84). His interpretation of the results of his experiments performed on hybrid peas lay forgotten for half a century, but finally paved the way for the science of genetics

Mendel observed that when he crossed plants which were grown from round seed with those grown from wrinkled seed, or tall plants with short plants, then in the next generation the hybrid plants were usually like one or other of the parents. The hybrid plants were not intermediate in character between the two parents, as most biologists in his day imagined they would be. When tall and short plants were crossed, the offspring were tall; when round and wrinkled peas were crossed, the next generation had round seeds. Mendel developed the idea that the tall and round characters were 'dominant' over the short and wrinkled characters in the hybrids formed in the first generation, and that the 'recessive' characters – short and wrinkled – were somehow present in the hybrids although unexpressed: they were masked somehow by the presence of the dominant characters. He deduced that this was so because, when the hybrids were fertilized with their own pollen, the recessive – grandparental – characters reappeared in the progeny. For example, when tall hybrids were crossed with other tall hybrids the seed nevertheless yielded a proportion of small plants.

Mendel observed that the original characters reappeared in the second generation without any essential alteration and without any transitional forms appearing. He put it something like this: 'There reappear [in the second generation], together with the dominant characters, also the recessive ones with their peculiarities fully developed; and this occurs in the definite average proportions of three to one.' For example, in one experiment Mendel found that on crossing round pea hybrids he obtained 5,474 round peas in the offspring to 1,850 wrinkled ones, a ratio of 2·96 to one.

Mendel went on to observe that the plants which showed recessive characters in the second generation did not show any further variation when self-fertilized. However, when the plants possessing the dominant characters were self-fertilized he found that some plants gave offspring showing the recessive characters, while some did not. In fact, he found that two-thirds of the plants possessing the dominant character yielded, on self-fertilization, offspring displaying both dominant and recessive characters, while one-third of the plants yielded offspring which did not vary. He concluded that in the offspring of self-fertilized tall or round hybrids there was an overall ratio of one true-breeding tall and round-seeded plant to two hybrid tall and round-seeded plants to one true-breeding short and wrinkled-seeded plant.

Mendel deduced from these experiments, and from other experiments in which he crossed plants differing in two or more characters, that there are hereditary factors, now called genes, in both the egg and pollen cells. The proportions of these genes in the egg and pollen cells, he believed, accounted for the proportions of the different kinds of plant in the progeny. In fact, he deduced that there were two genes – one from the father and one from the mother – in the cells of the plant. Mendel developed this idea mathematically and showed that the genes he studied behaved independently of each other and according to simple laws of probability. He did not know that the genes which he studied were located on the chromosomes.

We now understand that diploid organisms such as peas inherit one chromosome from the father and the other one from the mother for each pair of homologous chromosomes. If different genes are inherited from the father and mother, then usually one gene will prove to be dominant, although sometimes the hybrids may have a form intermediate between that of the dominant and recessive forms of the parents. If the effects of the genes are analysed biochemically, the effects of both dominant and recessive genes can often be detected. In the case of sickle-cell anaemia, an inherited blood disease in man, the anaemia gene is recessive but can nevertheless be detected by special tests in hybrid carriers of the gene.

Mendel originated a terminology for describing the behaviour of characters in crosses which is still used, with only a few modifications, today. If a plant is a pure-breeding tall variety it can be designated *TT*, showing that it has inherited the dominant gene for tallness from both its parents. Similarly, a short plant would be designated *tt*. The hybrids of a cross between tall and short parents would be designated *Tt*, showing that although these plants are tall they have the recessive gene for shortness as well. The combination of letters used to describe an individual is called the genotype of that individual.

If a hybrid plant *Tt* makes eggs or pollen, the sexual reduction divisions occur so that eggs and pollen with the constitution *T* or *t* are produced in equal numbers. If such a hybrid plant is self-fertilized, the egg and pollen cells fuse at random to give *TT* (tall), *tt* (short), and *tT* or *Tt* (tall hybrid) individuals in equal average numbers. Since *tT* and *Tt* individuals are the same there are on average two of each hybrid plant to each tall and each short plant.

Genes located far apart on the same chromosome or on different chromosomes are inherited independently. Genes located on the same chromosome are described as being 'linked together', and the closer together they are the more likely they are to go into the same egg or pollen cell. However, the process of crossing-over of chromosomes, which takes place during sexual cell division, may cause two genes on the same chromosome to become separated – so that one gene may lie on one chromosome of a homologous pair and the other gene on the other chromosome. These effects of crossing-over are used as a basis for making genetic maps of chromosomes, showing the positions of genes relative to one another along a given chromosome. These maps may often be very detailed – it is even possible to map differences on the DNA strands which lie within one gene.

Sex determination

In most organisms equal numbers of males and females are produced in the offspring – and this is ensured by a neat genetic mechanism. In humans there are 46 chromosomes, or looked at another way, 23 pairs. In females there are 22 ordinary chromosome pairs and one pair of sex chromosomes – the X chromosomes. Males also have 22 ordinary chromosome pairs, but they have only one X chromosome paired with a special male Y chromosome. It is this Y chromosome which makes a man what he is in genetic terms. Since the X and Y chromosomes are easily recognized under the microscope, sex can easily be established by carefully examining a single cell. In fact, this method has been used to foretell the sex of unborn babies by removing a few cells from the fluid surrounding the developing foetus.

Each egg produced by a human female contains an X chromosome. In the male, however, sperms of two kinds are produced – those carrying an X chromosome and those carrying a Y chromosome. The X-carrying sperms fertilize eggs to produce XX, or female, offspring, while the Y-carrying sperms fertilize eggs to produce XY, or male, offspring. X and Y sperms are produced in equal numbers and thus approximately equal numbers of boys and girls are conceived, although in fact an excess of boys is born – about 106 for every 100 girls. Exactly how this difference comes about is not yet known. Girls, however, are generally healthier than boys and live longer, so in older age groups the ratio is reversed.

The system of sex determination found in humans is the commonest form of sex determination among animals. But in moths, reptiles and birds it is the female which has an XY constitution and so produces both X-bearing eggs and Y-bearing eggs. The males in these groups of animals are XX and so can produce only one kind of sperm. Some plants have distinct X and Y sex chromosomes and in some cases, for example the strawberry, the female is again the XY sex. Many plants, although showing sexual differentiation, lack both X and Y chromosomes.

Since the mid-1950s technological advances have made it possible to study human chromosomes in detail. These studies have added greatly to our knowledge of the way in which sex is determined in man, and have explained several puzzling clinical syndromes involving abnormal sex organs. Klinefelters syndrome, for example, is the name of a condition found in male patients who suffer from development of the breasts, small testes and failure of sperm formation; and sometimes people with this condition are also mentally retarded. Examination of the chromosomes of patients with Klinefelters syndrome has revealed one or more extra sex chromosomes present in their cells. Usually they have only one extra sex chromosome – that is, they are XXY. But some may be XXXY, and others even XXXXY. Extra chromosomes are due to a fault arising when the number of chromosomes in the cell is being reduced during the

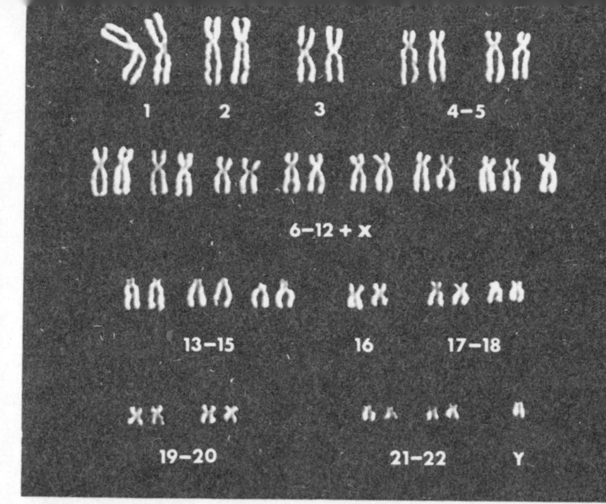

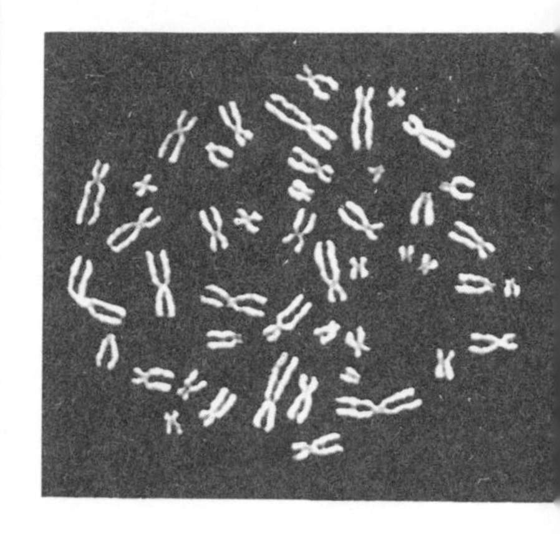

Human chromosomal abnormalities are most easily detected if the chromosomes in a micrograph of a single cell are cut out and arranged in a standard order. Here, a normal male chromosomal complement (above and above left) is compared with those of individuals suffering from Turner's Syndrome (Y chromosome missing) and Klinefelter's Syndrome (an additional X chromosome)

formation of eggs and sperms. These extra chromosomes are usually found in every cell of the body. However, sometimes 'mosaic' individuals are found who have a mixture of different kinds of cells in their bodies – perhaps a mixture of XY and XXY cells. Whether such a person is sexually abnormal or not depends on the proportion of the two kinds of cell present.

Another syndrome, known as Turner's syndrome, has been found in patients who have only a single X chromosome and no Y chromosome. Such individuals are described as XO. They have rudimentary female sex organs, short stature and a short webbed neck. This shows that in man the Y chromosome is necessary for normal male development – that it is not the presence of one rather than two X chromosomes, but the presence of the Y chromosome alone, that causes the cells to develop into a male. Normal fertile female development, however, is only possible where two X chromosomes are present.

This system of sex determination by no means applies to all animals. In the fruit fly *Drosophila* sex is determined by the ratio of X chromosomes to the non-sexual, or 'somatic', chromosomes. The male fly is normally XY, but XO flies, although sterile, are also normal-looking males. Normal female flies are XX, but XX flies with twice the normal number of the somatic chromosomes are males. Yet flies which have twice the number of all the chromosomes, both sex and somatic, are females, provided there is no Y chromosome present. The Y chromosome is, however, necessary for normal sperm development and fertility, so it cannot be dispensed with altogether.

There are several other interesting syndromes involving sex chromosomes in man. Males with extra Y chromosomes have been found in unexpectedly large numbers among the inmates of certain special mental institutions. These XYY individuals have generally been found to be mentally retarded; they also tend to be unusually tall and to have aggressive tendencies. However, quite normal XYY males have been found in the general population – so the effects of an extra Y chromosome are not invariable. But it seems quite safe to say that normal male aggressiveness has something to do with the presence of a Y chromosome in the cells of the male body.

Since normal women possess no Y chromosomes at all, one would expect the issue of a virgin birth – that is, a child conceived without male participation – to be female. The chromosomes in the unfertilized egg would have to double and this would simply result in an XX egg. Even if one of the X chromosomes were to be lost, the egg would be XO and so develop into a child with Turner's syndrome. Theoretically, however, it is not wholly impossible for a female to bear a male child by virgin birth, although there is no convincing evidence that this has ever actually occurred. Individuals have recently been found who are mosaics of XY and XX cells. It seems possible that an individual of this kind might be a female but nevertheless have some XY cells in the ovary or neighbouring tissues. It is possible that a normal X egg in such an individual might fuse with a Y-carrying cell to give an XY 'fertilized' egg cell, which would of course develop as a male.

Not all XY individuals are males: some are attractive females. There is a syndrome known as 'testicular feminization' in which XY individuals grow up as slim hipped females with well developed breasts. However, they have a small or rudimentary vagina with a blind end and scanty hair in the armpits and pubic region. These individuals have testes which secrete male hormones, but they nevertheless develop in the female direction because their cells are insensitive to the hormone. They are perfectly normal mentally, prove to be excellent mothers when they adopt children, and are often outstanding women athletes. Usually the testes are removed after puberty and they live their lives as females, although they can never have any children.

Mongolism is another condition caused by abnormal chromosomes. Those affected have abnormalities of the eyelid which give them an Asiatic appearance. This syndrome is also characterized by mental retardation and abnormalities of the face, tongue and other parts of the body. About 0·15 per cent of all children born are Mongols although older mothers give birth to a higher proportion than younger mothers. The majority of Mongols have been found to have 47 chromosomes instead of 46, one of the small chromosomes being present in three copies instead of two. How the extra chromosome actually causes mongolism in biochemical terms is not known. However, as with the sex chromosomes, the balance or proportion of the different chromosomes present in the cell is important. It is not sufficient to have at least one copy of each gene present in the cell: they must be present in the correct proportion.

A variety of other types of chromosome abnormality also occur in man, but few survive birth. A large number of natural abortions (miscarriages) and stillbirths are found to have abnormal chromosomes, and it is surely a blessing that these children are not born alive.

THE CHEMISTRY OF LIFE 5

In 1812 chemists discovered that certain inorganic chemical reactions could be speeded up if small quantities of other substances were present; and the substances which speeded up the reactions were found to remain unchanged after the reaction was completed. A new process – catalysis – had been discovered. Cell function has since proved to be heavily dependent upon catalysis: the living cell could almost be thought of as a bag of special catalysts – the enzymes – arranged in such a way that it is able to grow and reproduce rapidly and efficiently. Most cells contain hundreds or thousands of different enzymes, which catalyse as many different reactions. Enzymes are involved in the reproduction of DNA and in the synthesis of RNA and protein; they are also necessary in the chemical reactions associated with the storage and utilization of energy by the chloroplasts and mitochondria, and with the breakdown of food substances in the stomachs and intestines of animals. These digestive enzymes are now familiar to everyone as a constituent of the 'biological' washing powders that 'digest away the dirt'. The enzymes put into these powders are in fact derived from the cells of bacteria.

The first enzymes were discovered in 1814 in extracts of fermented barley which were found to have the ability to break down starch into glucose. Ferments, as enzymes were then called, were also recognized as playing a part in the processes of digestion. In fact, saliva was found to contain, in common with the barley extracts, an enzyme, called diastase, able to break down starch into glucose. The significance of enzymes as a general phenomenon was not however realized until the German biochemist Eduard Buchner discovered in 1897 that extracts of yeast cells containing no whole cells could ferment sugar. Buchner had been studying the proteins of yeast, and had made extracts which he tried to preserve, as fruit is preserved by the housewife, by addition of sugar. To Buchner's surprise the sugar was converted into alcohol and carbon dioxide by fermentation – and so began one of the major activities of modern biochemistry: the study of enzymes and their reactions.

All enzymes are now known to be proteins which are coded for by the DNA of the cell and synthesized in the cytoplasm. Proteins were at first thought by chemists to be mixtures which could never be resolved into their components; it was considered quite futile to attempt to purify an enzyme. However, in 1926 in the United States, J. B. Sumner succeeded in crystallizing the enzyme urease, which he extracted from Jack beans. Since then, very many enzymes have been purified – although it can take one scientist up to a year or more to find a method of purifying a single enzyme. The purification of enzymes has enabled biochemists to simulate many biochemical pathways in the test tube and so to verify and extend their hypotheses. More recently it has even been possible to use purified enzymes to synthesize other enzymes with the help of RNA messenger molecules.

When an enzyme catalyses a reaction a substance known as the substrate is converted into the product of the reaction. Starch, for example, might be the substrate, and be converted by enzyme action into the product sugar. In a single minute, one enzyme molecule can convert hundreds, thousands, sometimes even millions, of molecules of substrate into product molecules. Enzyme molecules themselves are unchanged by the reactions they catalyse, although like all proteins they are susceptible to slow decay.

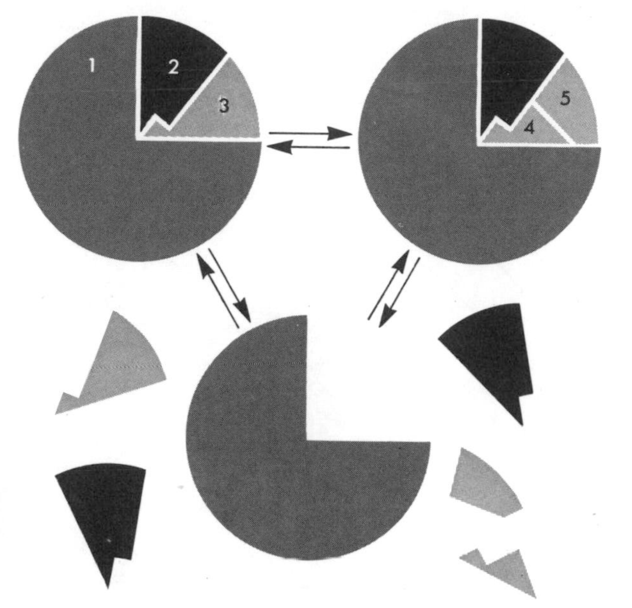

How an enzyme works. Enzyme molecule (1) has a cleft in its side which specifically accommodates an activator, or cofactor, molecule (2) and a substrate molecule (3). A strain causes the substrate molecule to split up into product molecules (4) and (5), which are then released. Enzyme reactions are frequently reversible

Some enzymes are located in particular cellular organelles, such as the mitochondria, chloroplasts, lysosomes and other cell organelles; but other enzymes are free in solution in the cell sap. In whatever way the enzymes are distributed in the cell their activities can be described in terms of the reaction pathways they form – the product of one enzyme reaction usually acting as the substrate for another, giving a complex network of interconnecting chemical reactions. A simple cell might easily have a thousand different enzymes linked up with each other in many different ways; here, however, we can only consider some of the more important pathways.

Not all pathways are found in all organisms. Organisms are often dependent on one another for their supplies of essential chemical substances and may utilize the waste products and dead parts of other organisms. Some of the solar energy that plants trap by photosynthesis is stolen by animals and micro-organisms; and plants in their turn are often dependent upon micro-organisms that convert atmospheric nitrogen into ammonia or nitrates, so fertilizing the soil. Besides the biochemical pathways within the

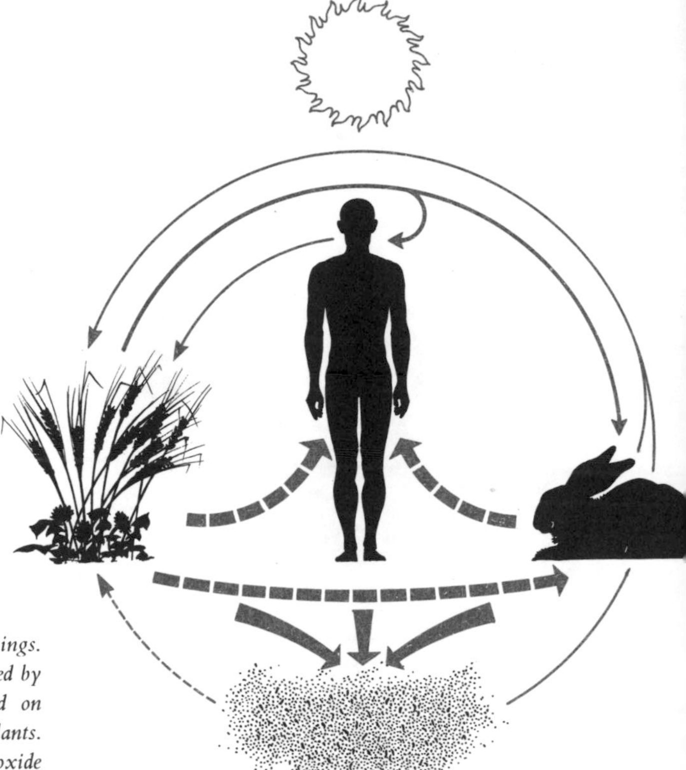

The interdependence of living things. Animals depend on food supplied by plants and other animals and on oxygen exhaled by green plants. Green plants utilize carbon dioxide exhaled by animals and are nourished by the products of animal and plant decay

organism there are biochemical cycles outside the organism which involve the interchange of carbon, nitrogen, sulphur and other chemicals between one organism and another.

Energy from the sun – photosynthesis

Life on earth could not continue without the photosynthetic activity of plants. The ability to obtain energy from the sun by photosynthesis must have been one of the first biochemical processes to have evolved in primitive living things. By photosynthesis carbon dioxide and water are converted into larger molecules composed of carbon, hydrogen and oxygen. Several molecules of carbon dioxide combine with several molecules of water, and excess oxygen is liberated as a gas. That the liberated oxygen is derived from the water molecules and not from the carbon dioxide was shown conclusively in 1941 in one of the first experiments in which radioactive isotopes were used. If 'labelled' water made from the oxygen isotope ^{18}O is used by photosynthesizing cells, all the isotope is later recovered as oxygen gas – the hydrogen atoms break away from the water molecules and combine with the carbon dioxide to make sugar.

In bacterial photosynthesis other substances, such as hydrogen sulphide, are used as hydrogen donors instead of the water used by higher plants. The process involves the removal of hydrogen atoms from their very stable association with oxygen in water molecules (or with sulphur in the case of hydrogen sulphide), and the transfer of these hydrogen atoms into a less stable association with carbon molecules to form carbohydrates. Oxygen atoms left together by this process associate to form unstable oxygen gas. In this way energy is stored in the carbohydrates to be released once more when the carbohydrates combine with oxygen in the slow-burning process of cell respiration.

The whole process of photosynthesis occurs in the chloroplast. Recently it has been shown that chloroplasts isolated from the cell and placed in a test tube can be made to perform the whole photosynthetic process, if they are maintained in the right condition and supplied with the right chemicals. Very little is known

The sequence of events in photosynthesis. Light energy absorbed in chlorophyll molecules is used to boost the energy of hydrogen atoms, which finally combine with carbon dioxide to form various sugars

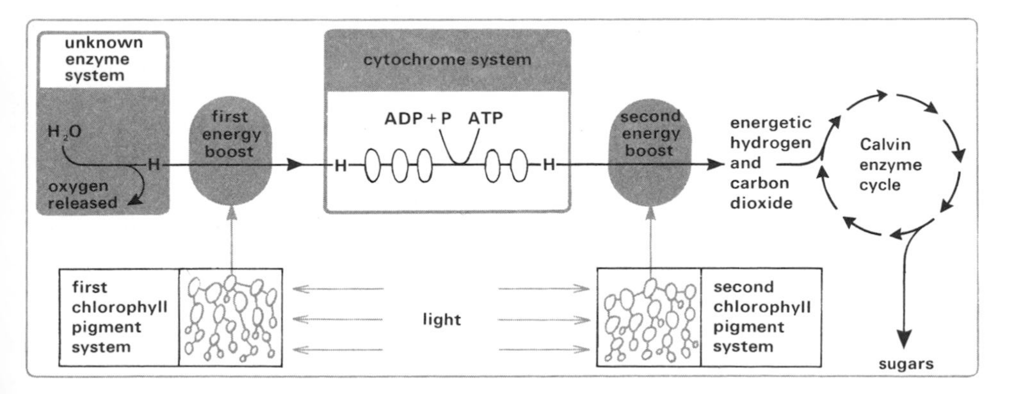

about how the first stage of photosynthesis works, except that a number of different enzymes are involved More is known about the second stage during which energy is collected by what have been called the 'photosynthetic units'. There are about three hundred chlorophyll molecules in each of these units. Each of the three hundred molecules absorbs light energy, which is then transferred from one molecule to another – the energy, as it were, travelling through the unit. Eventually all the energy reaches a special pigment molecule in which it is trapped. This trapped light energy reacts with hydrogen atoms, which have been separated by enzyme activity from water molecules, and boosts their energy. The resulting energetic hydrogen atoms may then combine with carbon dioxide in a complex process involving the enzymes of the Calvin cycle – until at last the energy originally absorbed in the form of light is temporarily stabilized in the form of sugar molecules.

Energy off the shelf – respiration

Just as photosynthesis may be thought of as the process of trapping and storing energy, so cellular respiration may be viewed as the process of mobilizing energy stores for use by the cell. In the course of this process the relatively complex molecules of sugar and starch are broken down into carbon dioxide and water, and it is in this way that the energy stored up by photosynthesis is made available.

In large animals the waste gases of cellular respiration are carried in the blood to lungs or gills, where they can be exchanged with the oxygen of the air. Most plants respire through pores in the leaf surfaces: during daylight hours they give out oxygen because their photosynthetic activity is greater than their respiratory activity, but at night little or no photosynthetic activity occurs, and the unused carbon dioxide is therefore given off into the atmosphere. When photosynthesis is occurring, the carbon dioxide produced by respiration is immediately recycled and converted once more into starch by the photosynthetic

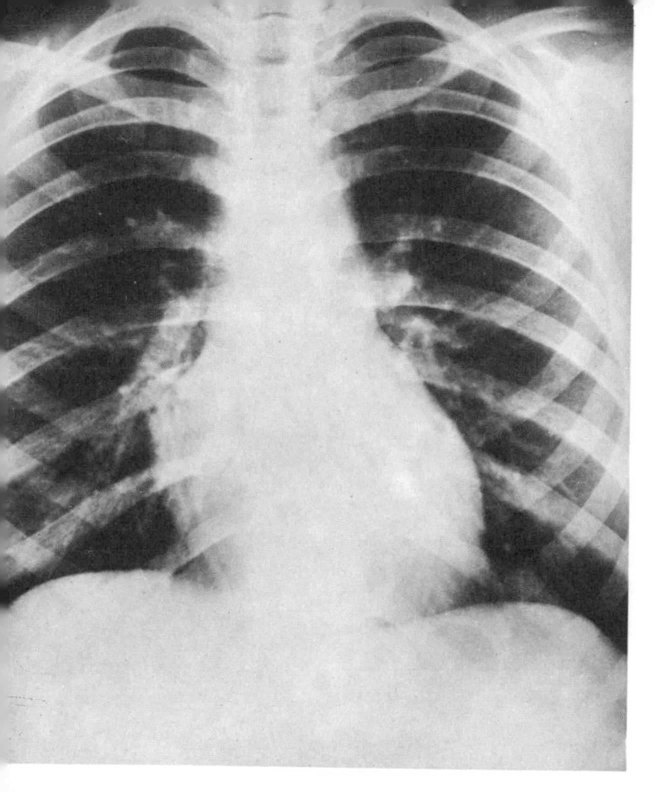

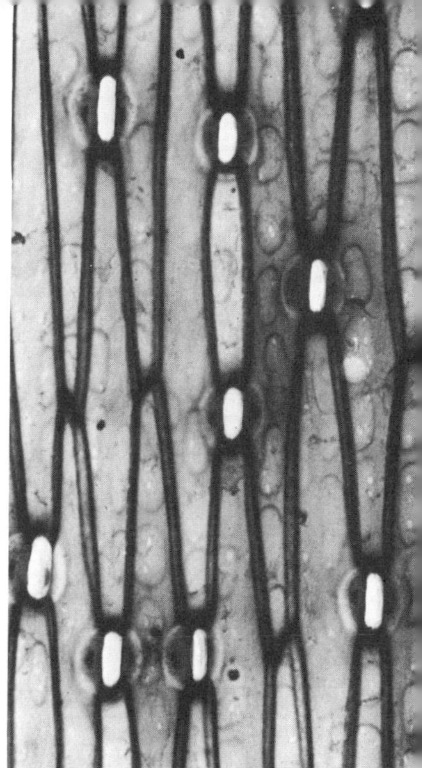

Before respiration can take place in cells it is necessary for the cells concerned to be immersed in a fluid containing free oxygen. Airborne oxygen diffuses into the human bloodstream through the delicate membranes of the lungs (X-rayed left); it enters the tissues of the iris leaf (above) through a myriad tiny pores

processes. Under optimal conditions photosynthesis can proceed at a rate as much as forty times more rapid than respiration.

Cell respiration may use up oxygen from the atmosphere (aerobic) or proceed without using oxygen (anaerobic), in which case lactic acid or alcohol are formed as the end products of respiration, rather than water. Thus yeast cells, in the absence of air, make alcohol. Anaerobic respiration releases only about one twentieth of the energy available in the glucose molecule and so is much less efficient than aerobic respiration.

Some organisms specialize in anaerobic growth and are able to make use of energy supplies in the absence of oxygen. There are some micro-organisms able to exist and multiply in the centres of heaps of decaying refuse where no air can penetrate or in the depths of stagnant ponds where there is no oxygen. Plant and animal cells are capable of living for short periods without air, but usually not for long. The muscle cells

As he breasts the tape at the end of a mile race, this world-class athlete is clearly feeling the effects of the oxygen 'debt' he has incurred

of an athlete build up an 'oxygen debt' during a sprint. Energy is available in the normal way from aerobic respiration, and the muscle cells are also able to obtain energy anaerobically with the build up of lactic acid – an organic molecule with a high proportion of hydrogen. But after a sprint fresh oxygen must be obtained during a rest period in order to burn up this lactic acid, thereby converting it into carbon dioxide and water. This also explains why the long distance runner cannot afford to build up an oxygen debt and accumulate lactic acid until near the end of his race. In a similar way yeast cells can grow either aerobically or anaerobically. In anaerobic growth yeast ferments sugar to produce alcohol. The process of fermentation is the same in both animals and plants, except that animals produce lactic acid rather than alcohol.

A complex enzyme pathway in the mitochondria is reponsible for slowly releasing the energy from the carbohydrate molecules as energetic hydrogen atoms. The energetic hydrogen atoms then use up their energy in forming ATP molecules (high-energy phosphate) and end up having spent their energy by forming alcohol, lactic acid or water molecules. Altogether 30 ATP molecules are formed from each sugar molecule which starts off. The process is calculated to be 60 per cent efficient – much more efficient than any man-made machine.

The high-energy ATP molecules supply energy to all parts of the cell so that all the many different organic molecules can be made. In plants these molecules are synthesized from salts, carbon dioxide and water, but in animals many molecules are received in food, and can be used with comparatively little change in their nature. Amino acids are synthesized in plants from nitrate salts; and these amino acids are in turn used to make proteins. Many other substances such as vitamins and fats are synthesized in the cell and are essential for normal cell activity.

The organic molecules used to build up the large RNA and DNA molecules must all be synthesized in a complicated series of steps with one enzyme catalysing each step. In the cell of a colon bacillus there are altogether probably about a thousand enzymes joined up in a tangled three-dimensional web of pathways which have not yet been fully explored. The flow of substances through the pathways of metabolism is maintained by an elaborate set of controls, so that the organism continues to grow despite a varying environment. The controls also prevent the pathways from becoming flooded or choked; and in this way the organism is able to respond to new conditions with sensitivity and precision.

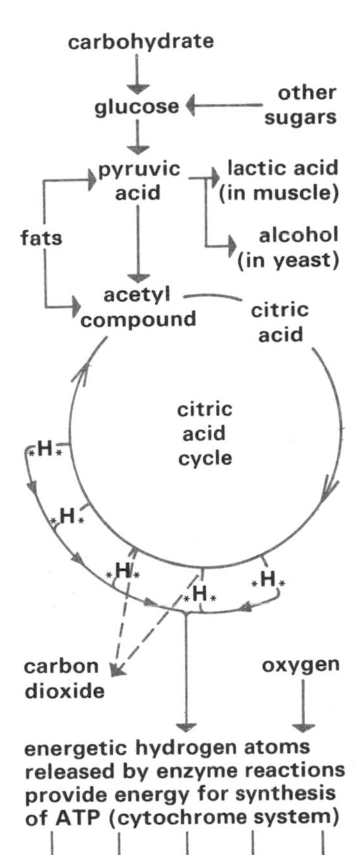

Respiration in the metabolic mill. Carbohydrates and sugars are burned, resulting in the release of energy in the form of ATP. Oxygen is consumed and carbon dioxide and water are produced

Feedback controls

Running the water for a hot bath is difficult to do without constant supervision – the water is likely to be too hot, too cold or too deep. The cell has the same problem in controlling the flow of substances from one enzyme reaction to the next along its various biochemical pathways. The flow is controlled by regulating the quantity and activity of the enzymes catalyzing the various reactions of the pathway. Some understanding of the way in which this type of control is achieved by cells came in the late 1950s through the work of three Frenchmen – Francois Jacob, Jacques Monod and André Lwoff. Their work developed rapidly, and in 1966 they achieved public recognition, being jointly awarded a Nobel prize.

Ten years ago little was known about the way in which cellular reactions were controlled. Today the mechanism controlling these reactions in micro-organisms has become fairly clear, and similar systems of control have also been found in the cells of higher organisms.

One of the commonest ways in which enzyme pathways are controlled seems to involve the product of the pathway inhibiting the activity of an enzyme which operates 'earlier' in the same pathway. For

Francois Jacob and Jacque Monod (right) who, with André Lwoff, gained a Nobel prize for elucidating metabolic control systems in the living cell

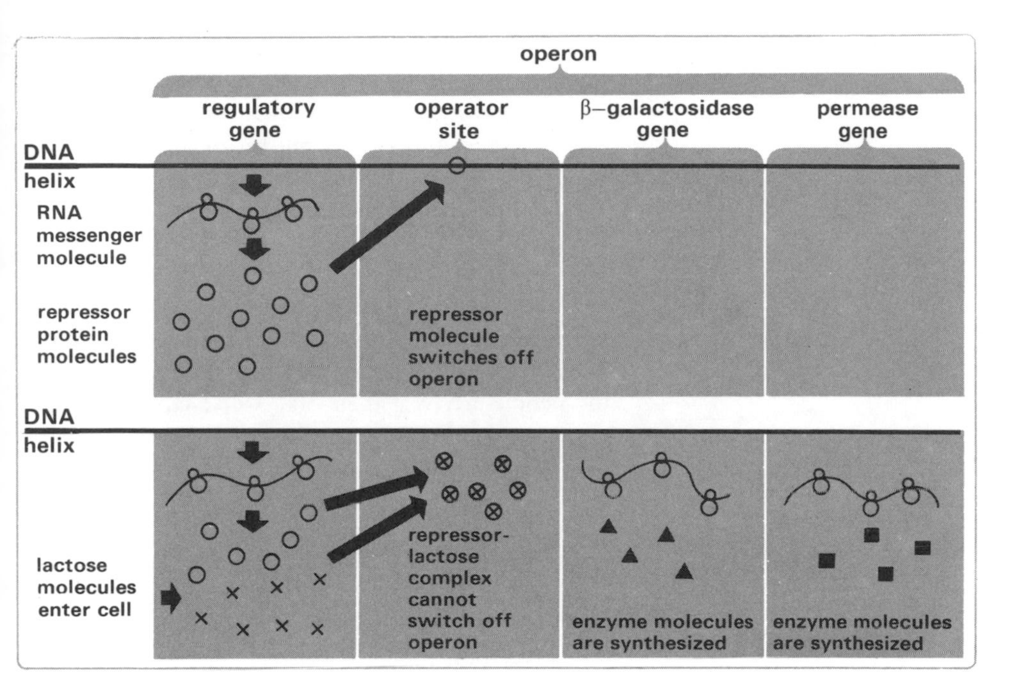

The operon-gene control system, discovered by Jacob and Monod in the colon bacillus. When this bacterium is in the presence of lactose sugar, genes are switched on to produce enzymes which cause sugar to be brought into the cell and to be broken down to make energy. In the absence of the sugar the genes are 'switched off', and these enzymes are not produced

example, the amino acid histidine will inhibit the catalytic action of an enzyme leading to its own synthesis – if histidine is present in the cell in a high enough concentration. In this way the amount of histidine synthesized is reduced, and its concentration in the cell falls until the enzyme 'switches on' once more and further histidine is made. This is a 'positive feedback' system – in principle the same sort of system used to regulate temperature in refrigerators and electric heaters. It also operates in the situation in which the more self-indulgent among us turn the hot tap on from time to time to keep the bathwater warm.

What actually happens is that histidine molecules combine with the enzyme molecules to produce a different 'switched off' enzyme molecule. The structure of the enzyme protein probably changes slightly so that the enzyme can no longer catalyse the reaction – the substrate molecule no longer fits neatly into the cleft in the protein. When the histidine concentration in the cell goes down, the histidine molecules break away from the enzyme molecule, which returns to its former shape and catalyses the reaction. While the enzyme is switched off the substances which come before it in the pathway may pile up, or, more likely, be diverted to some other pathway.

Imagine a branching network of roads and pathways radiating out from a railway station. There are gates at certain points – some open and some closed. People and goods in lorries are constantly leaving the station and walking along the various roads – appearing always to take the path of least resistance. Eventually the people arrive at their offices or factories and the goods are delivered. Imagine that the various gates and road crossings are controlled by policemen with walkie-talkie sets who receive their instructions from other policemen at a local headquarters. The policemen receive messages from the factory managers and control the flow of people and goods so that they arrive at the factories in suitable numbers. This imaginary police state is perhaps something like the cell. The flow of molecules through the biochemical pathways of the cell is regimented by control systems operated by the cell's genetic instructions.

Feedback control systems of the kind known to exist for histidine, and other amino acids are believed to exist for all the amino acids. These control systems serve to maintain suitable concentrations of amino acids in readiness for protein synthesis, which is going on continuously. The synthesis of DNA and RNA is controlled in a similar way. In fact there are at least ten control points in the pathways of RNA and DNA synthesis at which enzymes can be inhibited by feedback of product molecules.

This system of enzyme control can be very elaborate; and in multicellular organisms different tissues may have more than one enzyme, with different control characteristics, catalysing the same chemical reaction. The enzyme lactate dehydrogenase catalyses the conversion of pyruvic acid into lactic acid in animal cells during the production of short bursts of energy. The lactic dehydrogenase molecule consists of four

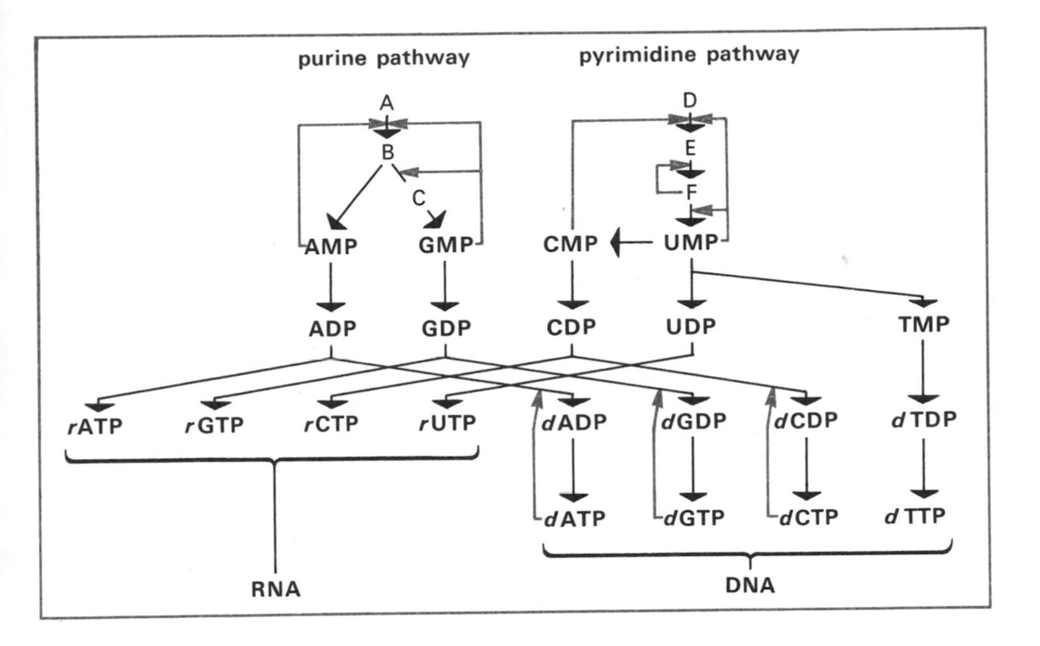

Diagrammatic representation of the control of DNA and RNA synthesis by feedback inhibition of enzyme activity. The control circuits are indicated in colour

subunits which can be of two kinds – H or M – in any proportion, as in HHHH, HHHM, HHMM, HMMM, MMMM. The heart muscle contains lactic dehydrogenase with a high proportion of H subunits, whereas other muscles contain lactic dehydrogenase with a high proportion of M subunits. These two varieties of the same enzyme have evolved to suit the different functions of the heart and other muscles. The heart beats continuously and the H form of lactic dehydrogenase responds to controls in such a way that little lactic acid is formed; whereas in other muscles, containing mostly the M form of lactic dehydrogenase, lactic acid may be formed in larger quantities during strenuous exercise and reduced in quantity during a subsequent rest period. The heart can never rest and so must not allow the lactic acid concentration to rise above a certain level.

The output of a biochemical pathway may also be controlled by the actual concentration in the cell of the enzymes catalyzing the reactions in the pathway. This was first established by experiments of Jacob and

Monod performed at the Pasteur Institute in Paris. They exploited the fact that the colon bacillus can grow using one or more of several different sugars as a source of energy. If cells of this bacillus which have been growing on glucose sugar are transferred to another sugar, such as lactose, then there will be a lag before growth can continue. During this lag the cell makes a new enzyme which can break down the lactose molecule. The instructions for making this enzyme exist in the DNA of the cell all the time, but they are only used, and the enzyme made, when the cell is in the presence of lactose. The enzyme in question, ß-galactosidase, breaks down the lactose molecule into two smaller sugars – glucose and galactose – which can then be used by the organism in the normal way. Another enzyme, also made by the cell at the same time as ß-galactosidase, is lactose permease, which assists the passage of the lactose molecule through the cell membrane. If the cells are transferred back to a glucose medium, they will immediately start using the glucose and, after a short lag, stop making both the permease and the galactosidase. If the cells are transferred to a medium which contains a mixture of glucose and lactose, then they use up the glucose first before adapting to the lactose by synthesizing the galactosidase and permease enzymes.

Enzymes of this kind, which can be switched on and off, are called adaptive enzymes. Now, it is known that very many enzymes are adaptive in the sense that the rate of their synthesis is adjusted to the needs of the cell. Small numbers of adaptive enzymes have been found to be organized together in groups, called operons by Jacob and Monod, which are controlled by another 'regulatory' gene. The regulatory gene is in fact an on-off switch – it produces a protein, called the repressor, which attaches itself to the DNA at a site called the operator and stops the synthesis of the RNA messenger molecules on the galactosidase and permease genes. However, if lactose is present in the cell, then lactose molecules react with the repressor protein and prevent the repressor from attaching itself to the operator site – and the galactosidase and permease RNA messengers can then be synthesized. Each of the genes – the galactosidase gene, the permease gene, the regulator gene, and the operator site – can be separately affected by mutation. If either the regulatory gene or the operator site are damaged by mutation, then the genes for making the galactosidase and the permease can become permanently switched on. If the galactosidase or permease gene is damaged, then only the damaged gene is unable to make its product in the normal way; the other gene and the control system are unaffected.

Many more operons have been discovered in a variety of different biochemical pathways in various micro-organisms, but not yet in multicellular organisms. Some operons have as many as six or more enzymes, all controlled as one unit. This type of operon control system is slower to react than the feedback-inhibition type of control. The two systems act in a complementary fashion – feedback inhibition gives a swift response and enzyme adaptation follows later. Enzyme adaptation prevents the wasteful synthesis of protein and allows organisms to keep genetic information in reserve – switched off until some change in the environment calls for its use. Reserve genetic information can be thought of as a secret weapon which the cell can use to give a flexible response to the environment. As we shall see in chapter 7, single-celled organisms have evolved ways of using these systems to exploit an extraordinarily wide range of environments.

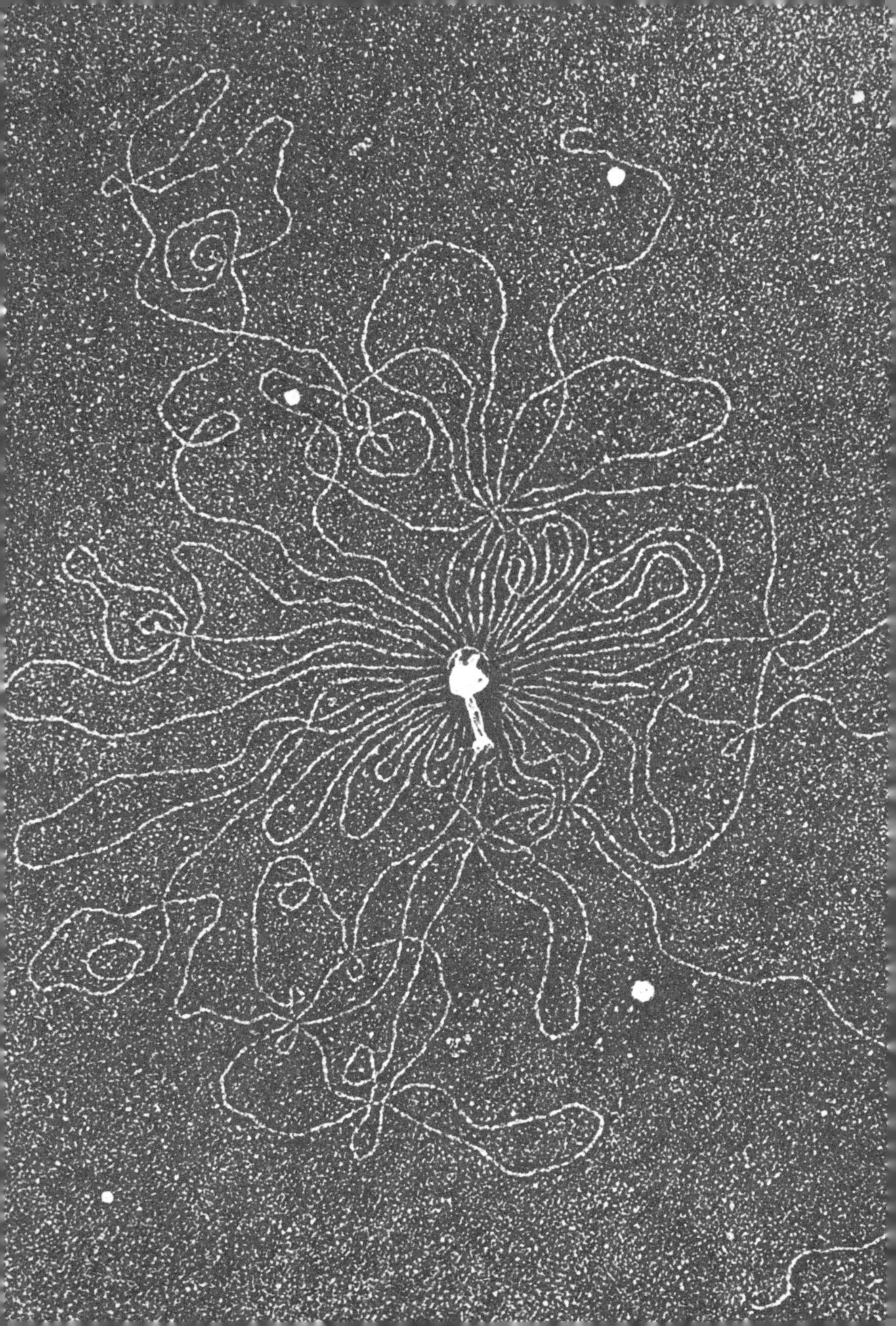

THE PROTOPLASMIC MACHINERY 6

In 1869 a young Swiss biochemist, Friedrich Miescher, was studying pus cells which he obtained by rinsing out bandages taken from hospital surgical wards. He tried to extract the nuclei from the pus cells and in so doing obtained a new substance, which he called nucleic acid. Miescher continued to study the substance over a period of several years, finding a rich supply of it in the sperm of the Rhine salmon – but nucleic acid turned out to be a complicated substance – too complicated for nineteenth-century chemists – and his work was put aside and forgotten for eighty years.

Nucleic acids – the master molecules

Interest in nucleic acid eventually revived as a result of some bizarre experiments, which were at first written off by many scientists as being merely sloppy work. These experiments were, however, to prove of the highest significance, leading to a revolution in biology and to fundamental changes in our knowledge of the nature of living things. In the magnitude of their effect on existing knowledge these changes can only be compared with those following upon the publication of Darwin's *Origin of Species*.

About forty years ago, Fred Griffith, an English bacteriologist, was experimenting with two different strains of bacteria – both causing pneumonia in mice – and was puzzled to discover that mice injected with a live strain of mild-pneumonia bacteria together with a different strain of heat-killed virulent-pneumonia

Opposite: the relatively enormous length of a DNA molecule is displayed in this astonishing electron micrograph. A bacteriophage–a virus that attacks bacteria – has burst as a result of immersion in water, discharging its DNA content. The magnification is 90,000 times

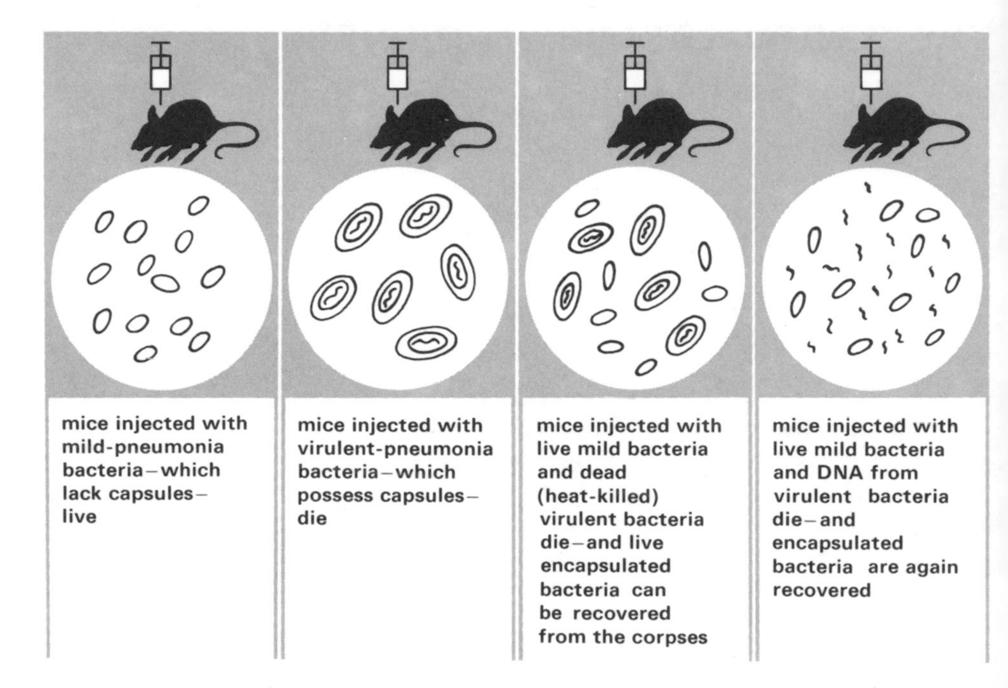

Griffith's experiment – one of the first to draw scientific attention to the remarkable properties of DNA

bacteria died of a generalized infection. Control experiments, in which mice were injected with either the mild strain or the heat-killed virulent strain alone, resulted in no serious infections. Somehow the heat-killed virulent bacteria were revived in the bodies of the mice by the presence of the live mild strain.

Now, the virulent bacteria possessed an outer cell wall – the capsule – which somehow enabled them to cause a lethal infection. The mild strain, however, was unable to form such a capsule. Yet Griffith was able to show that live virulent bacteria with capsules could be recovered from the mice at the end of the experiment. He suggested that the heat-killed virulent bacteria, rather than being revived by the live bacteria, somehow transformed the live bacteria into the virulent capsuled form.

Oswald Avery, an American chemist, did not believe Griffith's experimental claims and set out to disprove them. But, to his surprise, he found that he was able to repeat Griffith's observations in the test tube, and went on to show that it was the nucleic acid from the virulent strain which had transformed the mild strain. Avery later showed that it was the pure deoxyribonucleic acid (DNA) which was able to transform the cells, and not the protein, ribonucleic acid or purified substance from the capsule.

These results were not immediately accepted by the scientific community, as it was widely believed that protein was the genetic material. However, as a result of this work, and its subsequent elaboration, what happens in the transformation of one bacterium by the DNA of another is now understood. The DNA of the virulent strain is unharmed by the heat treatment which kills the cells. Some of this 'virulent' DNA enters the cells of the mild strain and combines with its DNA. The 'virulent' DNA carries genes which provide the information for the synthesis of special enzymes, which are in turn able to synthesize a capsule and so convert the mild into a virulent strain.

Nucleic acid is found in both the nucleus and the cytoplasm of the cell. Most of the nucleic acid from the nucleus is deoxyribonucleic acid and most of the nucleic acid from the cytoplasm is ribonucleic acid. These two kinds of nucleic acid are chemically different and play quite distinct parts in the working of the cell. All the cell nuclei in the body normally contain the same amount of DNA, whereas the sperm and egg cells contain half this quantity of DNA. This was one of the first facts to suggest that DNA was an important constituent of chromosomes. RNA is found in the cytoplasm of all cells, but in greatest quantity in cells such as those of the pancreas, which make large quantities of protein for export to other parts of the body – and this was one of the first observations to suggest that RNA might have an important role in protein synthesis.

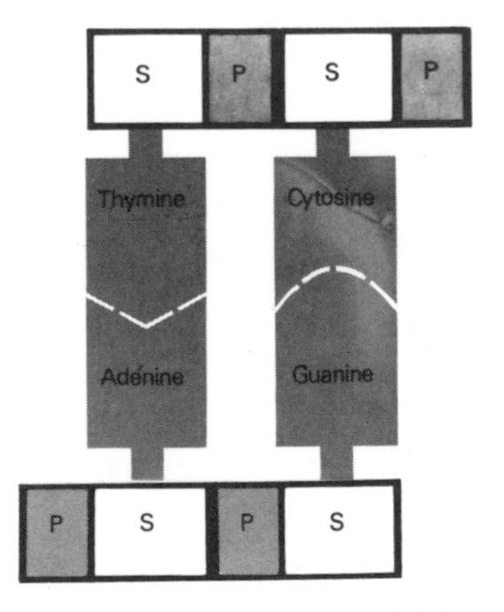

The basic structure of DNA. Twin chains of sugar and phosphate subunits are linked by 'rungs' comprising either the bases thymine and adenine, or the bases cytosine and guanine

Nucleic acids are constructed from four different types of molecular 'subunits'. The four DNA subunits are the bases adenine, guanine, cytosine and thymine. Each of these is attached in a specially determined order to a sugar-and-phosphate backbone to form a long polymer – which is a molecule containing repeating units. The sugar present in DNA is deoxyribose, while in RNA it is ribose. Another difference between DNA and RNA is that RNA does not contain the base thymine but contains another base instead – uracil. In 1949 the biochemist Erwin Chargaff, who was studying the chemistry of DNA, found that samples of DNA from different organisms contained the four bases in different proportions. However, he also found that the number of adenine and thymine bases was always equal (A = T) and so was the number of guanine and cytosine bases (G = C), but that the proportion A + T : G + C – called the base ratio – could vary from 0·3 to 3·0. These facts, as we shall see, were later found to be of the greatest significance.

All the biological evidence pointed to the central importance of DNA, and there accordingly developed something of a race to discover its structure. Clues came from a special technique – X-ray analysis – which had originally been developed for the study of much smaller molecules. There were immense problems in applying it to the study of large biological molecules. However, these problems were overcome and the result has been some of the most exciting and fundamental advances in modern biology. A beam of X-rays is shone through a crystal of the substance to be analysed and a photograph is taken of the pattern of spots produced by the bending of the X-rays by the crystal. After a complex mathematical analysis has been made of these spots, the shape of the molecules making up the crystal can be deduced and reconstructed as a scale model. The first detailed X-ray diffraction studies of DNA 'crystals' – actually they were fibres – were carried out in the London laboratory of Maurice Wilkins. Theoretical ideas about the actual structure of DNA were developed – in conjunction with Wilkins – by Francis Crick and James Watson, working in the Medical Research Council laboratories in Cambridge. The essentials of DNA structure were worked out in 1953, and all three men were subsequently awarded a Nobel prize. In the United States, Linus Pauling, who

won a Nobel prize for his work on proteins in 1954 (he also won the Nobel Peace Prize in 1963), was hot on the same trail. Pauling, however, proposed a structure for DNA which proved to be incorrect. Watson and Crick had spotted the error and published the correct interpretation before Pauling had time to change his mind.

The X-ray photographs showed – as Watson and Crick pointed out – that DNA was a coiled or helical structure. Watson and Crick then suggested that the

An X-ray diffraction study of DNA. It was from the analysis of evidence such as this that Crick and Watson were ultimately able to deduce the molecular structure of DNA

helix was double and that the bases, paired across from one strand to the next, were held together by attractive forces called hydrogen bonds, which gave the structure stability. Furthermore, they suggested that an adenine base should always pair with a thymine base and a guanine base with a cytosine base, so satisfying the rules about the ratios of bases derived from the chemical studies (A = T, G = C). However, they pointed out that any base could be followed by any other base along the strand, so allowing the freedom necessary to account for the variable base ratio A + T : G + C.

In 1953 Watson and Crick published a paper, in the periodical *Nature*, in which they dryly commented: 'It has not escaped our notice that the specific pairing we have postulated immediately suggests a possible copying mechanism for the genetic material.' This was all they had to say – and indeed perhaps all that could confidently be said at that time – about the biological implications of their work.

Discussing a DNA model in their Cambridge laboratory are Nobel prizewinners James Watson (left) and Francis Crick. A diagram showing how the DNA molecule reproduces itself appears on page 100

Before Watson and Crick discovered the structure of DNA, the way in which the genetic message was passed accurately on from one generation to the next had been a great mystery. But here now was a remarkably neat mechanism which made it all seem comparatively simple. The double strand of the DNA split down the middle and each half acted as a template for the synthesis of another half, the specific pairing of the bases working like a positive/negative printing process. This idea fitted in well with the theoretical ideas of geneticists, who were quick to seize upon it.

Enzyme structure

The incredible diversity of proteins and the obvious complexity of their structure led many chemists to believe that it would be impossible to analyse or understand them in any detail. Others saw little point in even trying to tackle such a problem. But Frederick Sanger, working in Cambridge, none the less set to work on the relatively small protein molecule insulin. Insulin is a readily available pure protein made from the pancreas of animals, and is familiar because of its use in the treatment of diabetes. Sanger used a method of analysis which involved first of all breaking down the protein sample into amino acids by a special acid treatment and then placing a sample on sheets of paper. He called this method 'fingerprinting' because a sweaty finger placed on the paper by accident leaves a print looking rather like a spot of amino acid. It took Sanger ten years of hard work to determine the sequence of amino acids in insulin. He completed the work in 1954.

Further experiments showed that the insulin molecule could be broken up by special treatments into two chains which were normally joined together by chemical bonds between two pairs of sulphur atoms. The separated chains were analysed: one was found to contain 20 amino acids and the other 31. The next stage in the analysis was to give the separate chains weak treatments with acid, so that they would break into pieces containing several amino acids, without being broken down entirely. These pieces, called peptides,

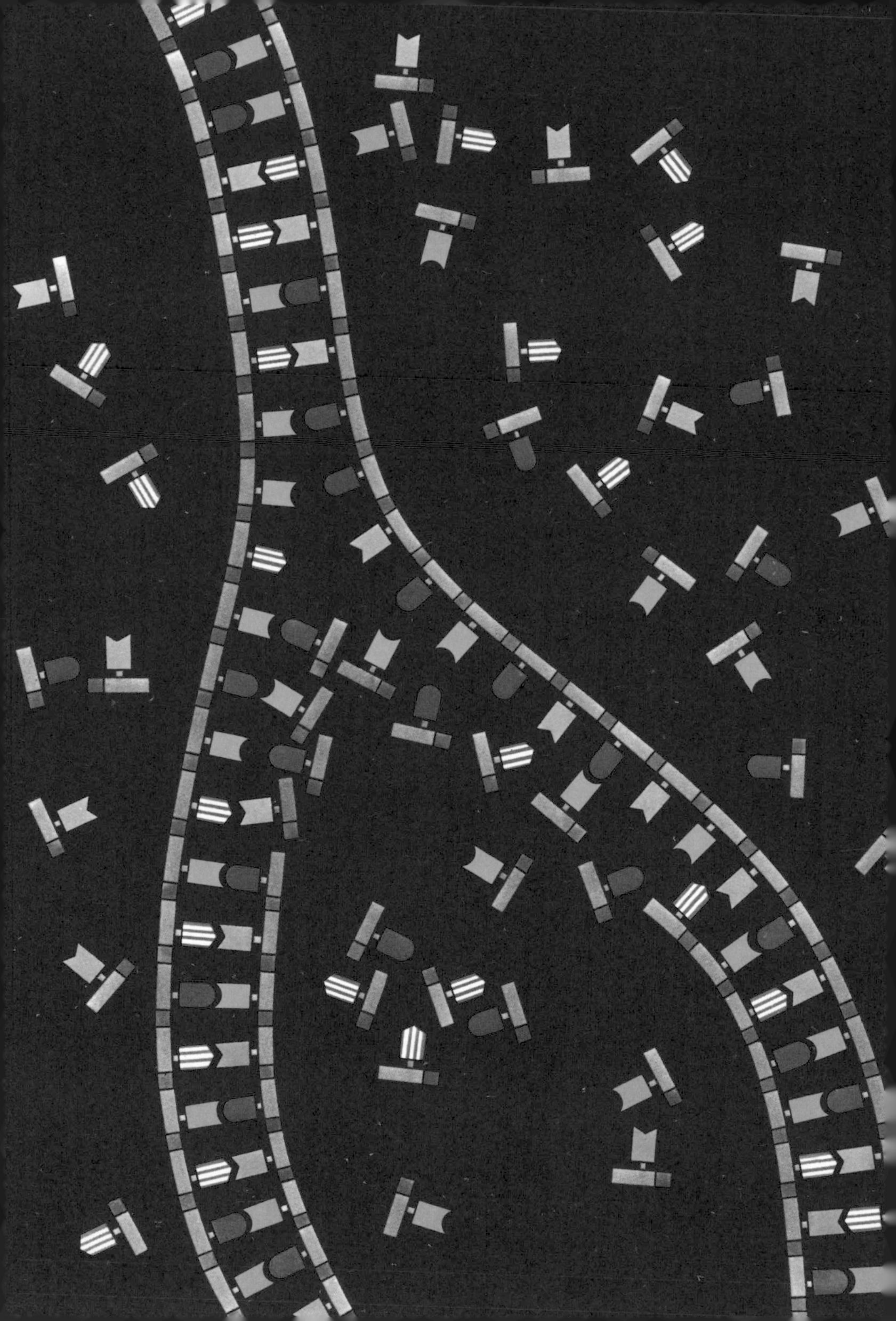

were then purified and the smaller among them analysed one by one to determine their amino-acid constitution. From the order of the amino acids in the smaller of these pieces the order in the larger peptide pieces could be established. However, a complete description of the chain was still not possible, since at some points along the chain's length breaks occurred very easily, and no overlaps were available. To complete the picture a further experiment had to be conducted in which the chain was broken by pure digestive enzymes obtained from the stomachs of animals. These enzymes broke the chain at points unaffected by the acid treatment and so gave a complete set of overlaps from which the whole molecule could be reconstructed. This type of approach has now been applied to a large number of proteins; and automatic equipment is now speeding up the analysis a great deal. The pioneer work which took Sanger ten years to complete could now be accomplished in as little as a year by an experienced investigator.

To determine the configuration and exact internal structure of protein molecules another Cambridge team, Max Perutz and John Kendrew, used the method of X-ray diffraction – a method first used with success on fibrous proteins, such as those of skin and hair. X-ray diffraction was not so easily applied to globular proteins such as enzymes, however. But eventually Kendrew was able to produce a model of myoglobin – a protein which carries oxygen in muscle – by combining the protein molecule with an atom of a heavy metal, which acted as a reference point to which the structure of the rest of the molecule could be related. Of the first results Perutz said: 'It was a triumph, yet it brought a tinge of disappointment. Could the search for truth really have revealed so hideous and visceral looking an object?' When analysed at a finer level, however, the visceral appearance of the myoglobin molecule became less obvious.

Shortly afterwards Perutz, using the same technique, successfully analysed the protein haemoglobin – the red pigment of blood which carries the oxygen from

Opposite: replication of the DNA molecule. The two chains forming the double helix separate, and each, combining with available free units, becomes the complementary template of a new chain and so of a new double helix. The four bases – thymine (striped red), adenine (green), cytosine (blue), and guanine (orange) play a crucial role is ensuring that an exact copy is formed

John Kendrew who, working at Cambridge in conjunction with Max Perutz, was able to determine the molecular structure of the protein myoglobin

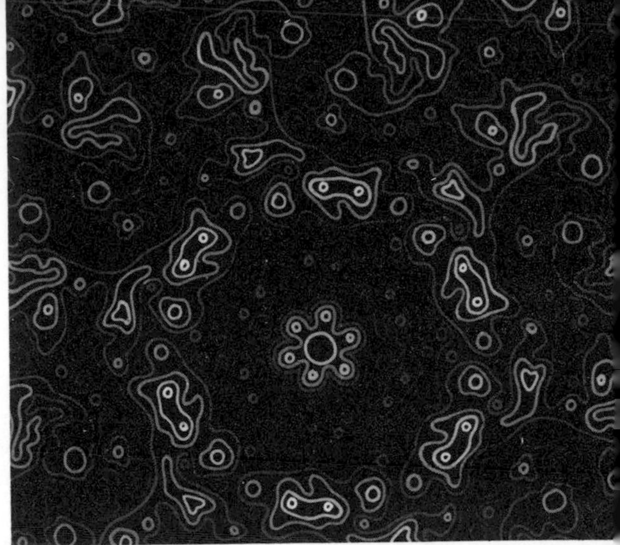

Below: a 'visceral' model of the myoglobin molecule, showing the convoluted path of the amino-acid chain. Right: an electron-density distribution 'map' of a 'section' through the insulin molecule

the lungs to the tissues. He found that haemoglobin consisted of units of two kinds – alpha and beta. Both the alpha and beta subunits of haemoglobin have a striking resemblance to myoglobin, so that the haemoglobin molecule, which consists of two alpha and two beta subunits, looks roughly like four myoglobin molecules stuck together. Perutz was also able to show how the haemoglobin molecule varies in shape depending on whether or not it is combined with oxygen. X-ray studies of crystals of haemoglobin in the presence and absence of oxygen showed that the two beta chains slide apart and lose contact with each other when not combined with oxygen, whereas the alpha chains remain unchanged. The two beta chains could be thought of as the jaws of a clamp which spring together and between them hold four molecules of oxygen. This was the first concrete proof that enzyme molecules – and haemoglobin is a special kind of enzyme – change shape when they combine with a substrate. Enzymes themselves could be thought of as little demons which wrap themselves round a molecule before reacting with it. Haemoglobin wraps itself round four oxygen atoms and carries them to parts of the body where oxygen is scarce; the oxygen then leaks off the haemoglobin to react with spent hydrogen atoms to give water. Other enzymes work in a similar way. Some enzymes wrap themselves round molecules and bend and twist them until they break; and some work by bringing molecules into a favourable position for reacting with one another in a way that would not otherwise be possible.

Studies of the amino-acid sequences of haemoglobin were first undertaken by Vernon Ingram, another Cambridge biochemist. The haemoglobin molecule consists of two identical halves, each made up of three hundred amino acids. The analysis of such a large molecule was an enormous task, so Ingram concentrated first of all on studying the differences between normal human haemoglobin and sickle-cell haemoglobin. Sickle cell is an abnormal haemoglobin responsible for a form of anaemia – characterized by the presence of sickle-shaped red blood cells – common in some parts of Africa. He found that the sickle-cell haemoglobin differed from normal haemoglobin by only one amino acid out of the three hundred. Sickle-cell anaemia was then called a 'molecular disease' – the fundamental cause of the anaemia being a defect in the haemoglobin molecules.

Sectional contour maps on perspex sheets are here superimposed to give a three-dimensional picture of the myoglobin molecule

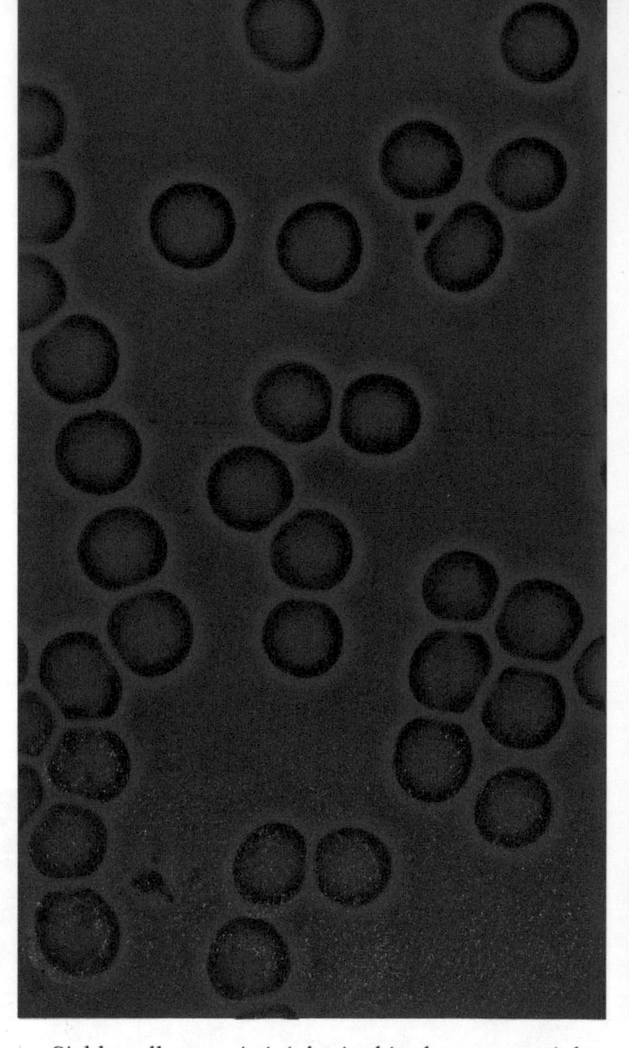

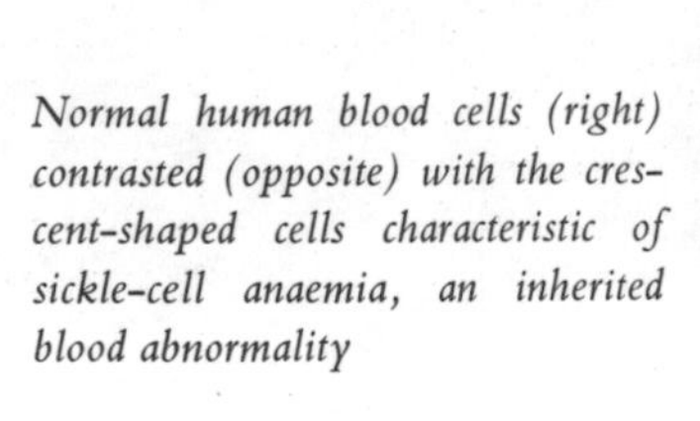

Normal human blood cells (right) contrasted (opposite) with the crescent-shaped cells characteristic of sickle-cell anaemia, an inherited blood abnormality

Sickle-cell anaemia is inherited in the same straightforward way as the genes investigated by Mendel which affect the height of peas or the roundness of their seeds. It was the first example to be analysed that showed how the genetic make-up of an organism exerts a fine control over the structure of its molecules (many other examples have since been found in organisms of all kinds). However, the sickle-cell anaemia is doubly interesting because although the disease results in the premature death of those who have two copies of the sickle-cell gene – one from the

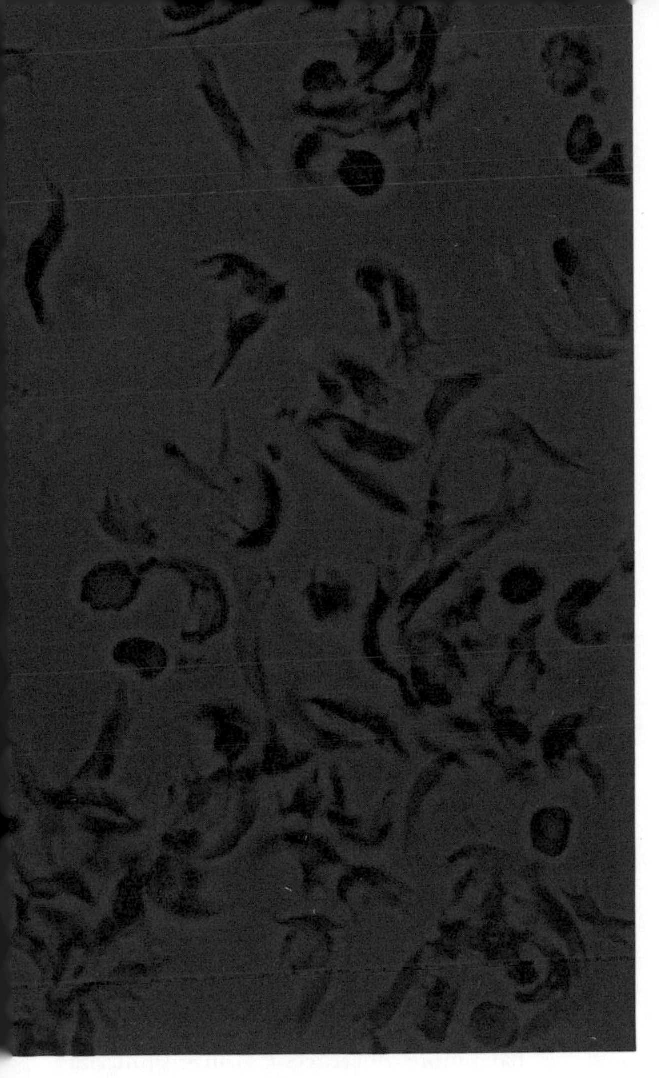

When sickle-cell anaemia is inherited from both parents, it is fatal; but if from only one parent, it is found to endow a resistance to malaria

father and one from the mother – it seems to confer an advantage on those who have only one copy of the gene by giving them a resistance to malaria. The sickle-cell disease is common in areas where malaria is common – the burden of the anaemia on the population apparently being counterbalanced by immunity to malaria. Perhaps after another few thousand years other gene mutations might occur which would reinforce the resistance to malaria that this gene confers, without causing anaemia when two copies of the gene are present.

Studies have now been made of the amino-acid sequences in many different haemoglobins from a large number of different animals. These have shown that about 22 of the amino acids in the alpha and beta chains of the horse, pig, cow and rabbit are different from those in the human alpha and beta chains. Assuming that in the course of their evolution these mammals and those in the line of human descent diverged roughly 80 million years ago, there is a net average rate of change of about one amino acid per chain (of 141 to 146 amino acids) per 7 million years. This is of course not simply the rate of mutation in the haemoglobin molecule, which would be much higher, but the frequency with which a successful mutation might be expected to establish itself in the population as preferable to the previously established condition.

There are several other haemoglobins found in man besides the alpha and beta types. These other types can be thought of as biochemical 'fossils', since they reveal how man has evolved at the level of protein structure. Gamma-chain haemoglobin is found in the blood of human and animal foetuses. It seems to serve a special function in the foetus, as it is replaced by beta chains after birth. Another type of haemoglobin – delta-chain – is present throughout life, but only in small quantities. The beta and delta chains in man closely resemble each other, there being only 10 amino-acid differences; the beta and gamma chains differ in the case of 39 amino acids. The alpha and beta chains, however, have more differences (77) than similarities (64). Myoglobin also shares similarities with the two haemoglobins in amino-acid sequence as well as in structural appearance. Comparisons of this kind lead to the conclusion that there was probably an ancestral haemoglobin chain from which all these other forms evolved. About 650 million years ago the ancestral haemoglobin gene must have accidentally doubled to form two genes which began to evolve separately, giving rise to the myoglobin gene and the haemoglobin gene. One of these genes must have doubled again about 380 million years ago – during the Devonian epoch, when the first amphibians were appearing – to give rise to the genes for the alpha chain and to another chain which was the ancestor of the beta, gamma and delta chains.

Protein production

When the cell grows and divides, not only does the DNA double in quantity, but everything else does too. But how does the cell grow in size, and how are the genetic messages of the DNA translated into protein, the principle substance of the cell? Cell growth is a complicated process and twenty years ago nothing at all was known about it. Yet the first experiments to find out whereabouts in the cell protein synthesis occurred were very simple. Animals were injected with a mixture of radioactive amino acids and shortly afterwards killed. Their tissues were analysed and the radioactive amino acids found to be attached to the ribosomes. These ribosome sites of protein synthesis are themselves half protein and half RNA. But there are also two other important types of RNA involved in protein synthesis, apart from the type of RNA that occurs in the ribosomes. These are: messenger RNA and transfer RNA.

The messenger RNA is made in the cell nucleus and is an exact copy of the DNA of one or more genes. Messenger RNA is actually synthesized, with the help of enzymes, using one strand of DNA as a template. This process, called transcription, can be reproduced in the test tube, using purified cell constituents. The resulting RNA is an exact replica of the DNA from which it is copied, although occasional copying mistakes can and do, of course, occur. The base ratio of RNA is the same as the DNA from which it is copied. Moreover, RNA copied from a particular DNA molecule can be shown to stick to it by a base pairing process similar to the process by which the two strands of DNA stick together: but RNA will not stick to DNA which has a different order of bases – even if the DNA has the same base ratio. There is a special equivalence between DNA and the particular

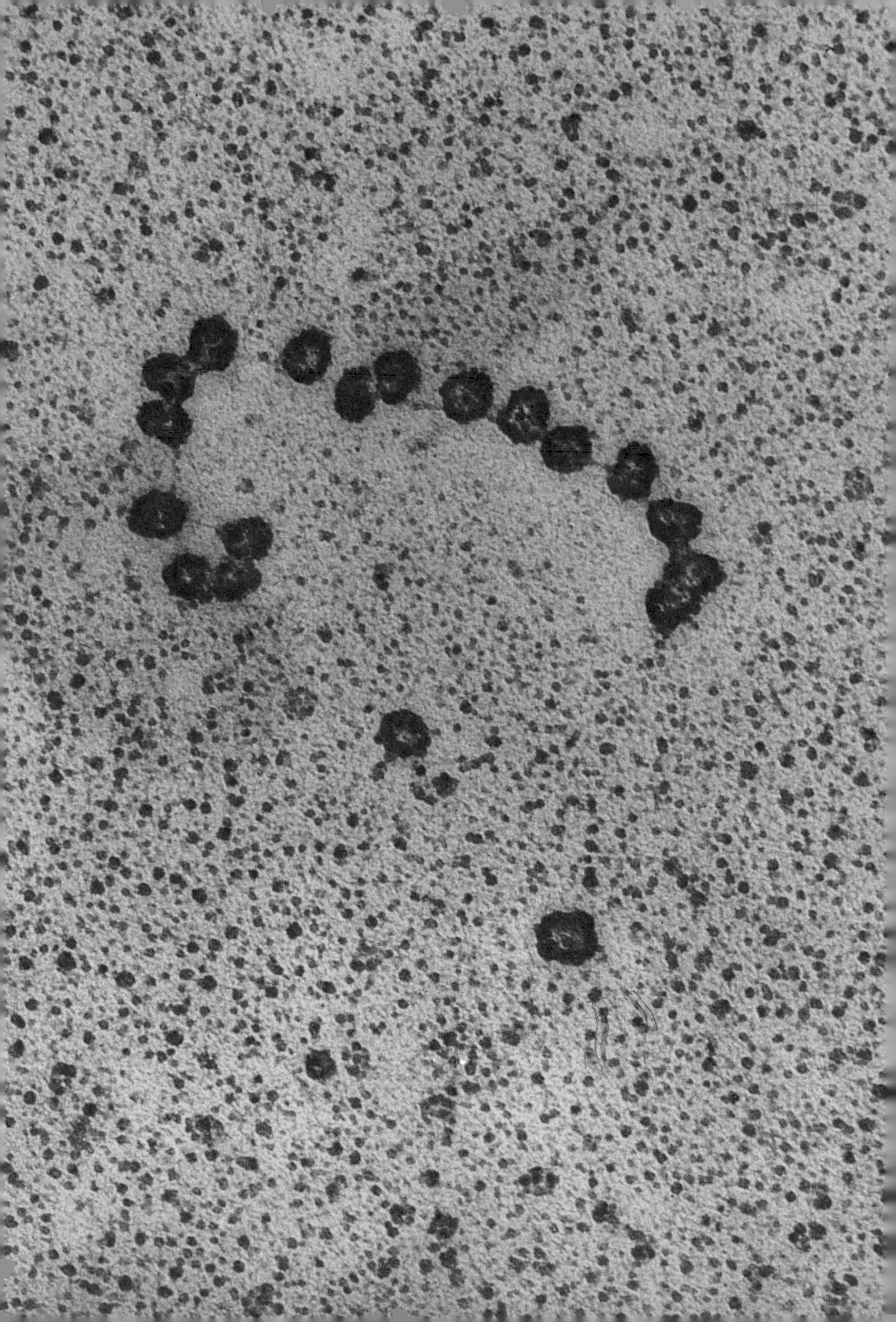

RNA molecule which is copied from it – and this is because the order of the bases along the RNA molecule corresponds exactly to that particular section of DNA.

Messenger RNA provides a means of getting the genetic message out of the nucleus and into the cytoplasm, and also a means of increasing the number of copies of the genetic message. Like a library with a copying machine, the cell is able to send out a large number of copies of selected stored information, yet still keep the original copy safely on the shelf. In this way the cell is able to respond to sudden demands for the product of a particular gene. When the messenger molecule has passed into the cytoplasm it becomes attached to the ribosomes, and at any one time several ribosomes may be attached to one messenger molecule, shuttling back and forth making protein molecules. This composite structure, which is called a polysome, is the machine that actually makes the proteins.

The four-letter code of the messenger-RNA molecules is translated into an amino-acid sequence by means of transfer RNA – the third kind of RNA found in the cell. The messenger-RNA molecules get the message which they embody turned into protein

The ribosomes apparently strung like beads along a messenger-RNA molecule (opposite) are sites of protein synthesis. They act as mobile workshops in which transfer-RNA molecules, arriving in a sequence dictated by the structure of the messenger-RNA molecule, deliver amino-acid molecules to the growing protein chain, as illustrated in the diagram (below)

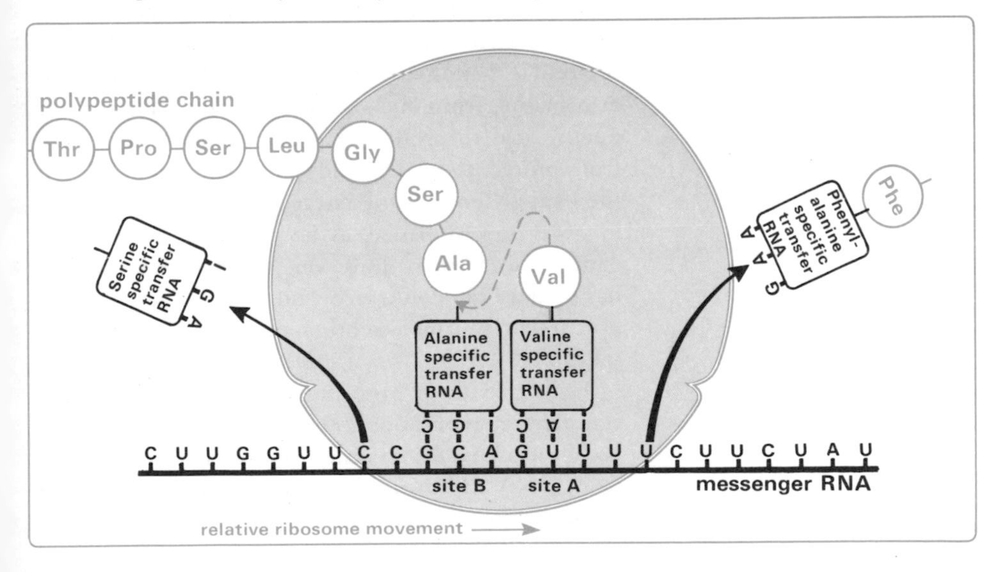

by using the transfer-RNA molecules as a sort of decoding device. Transfer-RNA molecules have the ability to attach themselves to an amino acid at one end and to a messenger molecule at the other. But each transfer molecule is capable of forming a link between only one kind of amino acid and the special 'code word' for that particular amino acid embodied in messenger RNA. When the messenger-RNA molecule has become attached to a ribosome, the transfer-RNA molecules come with their captive amino-acids and fit against the messenger molecule one by one. A base pairing process ensures exact recognition between the code word on the messenger RNA and the transfer RNA. While two amino acids are held together by the transfer RNA and positioned on the ribosome, an enzyme joins them together. The transfer RNA is released from the first amino acid and the messenger molecule moves along the ribosome a little so that the next transfer RNA plus amino acid can get into position. The process is repeated until all the amino acids are joined together into a completed chain which is then released as free protein. The free protein chain then spontaneously coils up into its characteristically convoluted shape.

The essentials of protein synthesis are summed up in what Crick has called the 'Central Dogma' of molecular biology. This states that 'the transfer of information from nucleic acid to nucleic acid, or from nucleic acid to protein may be possible, but transfer from protein to nucleic acid is impossible'. Crick gave the name 'Central Dogma' to this statement in 1958 in order to emphasize, as he put it, its 'speculative nature'. Since then, however, the 'Central Dogma' has been amply confirmed, and it is now usually forgotten that its name was originally given it as a donnish joke.

The three basic processes of genetic information transfer are: replication, transcription and translation. Replication is the transfer of genetic information from DNA to DNA when cells divide; transcription is the transfer of genetic information from DNA to

RNA; and translation is the transfer of genetic information from RNA to protein. The flow of information, as the Central Dogma states, is always from nucleic acid to nucleic acid or from nucleic acid to protein.

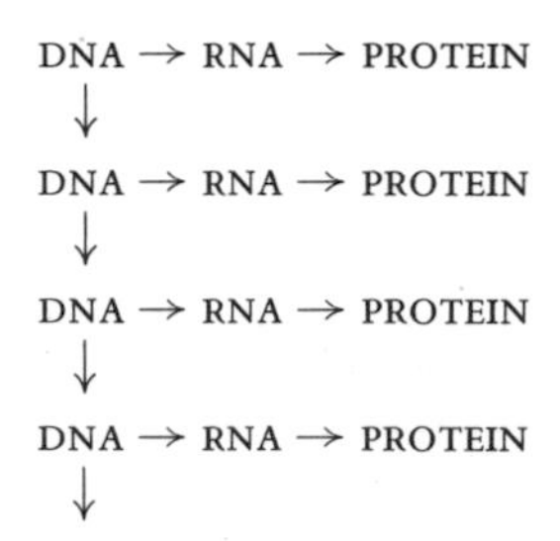

This is how the genetic information is passed on – but how is it stored, or encoded, in the nucleic-acid molecules? Egyptologists were fortunate in having the Rosetta Stone to help them decipher ancient Egyptian hieroglyphics. But molecular biologists could look to no simple key to the genetic code. Instead they had to develop powerful experimental techniques to do the same job.

The genetic code

There was much initial speculation about what kind of code might be hidden in the structures of nucleic acids. There were known to be four different bases – an 'alphabet' of four 'letters' – to code for twenty different amino acids. But the size of the code words and the rules by which they worked remained a mystery – at least until 1965.

As is often the case in science a number of gifted men were to contribute to the code's solution. In the United States Marshall Nirenberg developed a test-tube 'system' containing ribosomes, transfer RNA, amino acids and other components involved in the synthesis of proteins; and in 1955 a Spanish-American biochemist, Severo Ochoa, discovered an enzyme which would join bases to form messenger molecules in a test-tube. Nirenberg then found that artificial messengers made by Ochoa's method could act as templates in his test-tube system and so direct the synthesis of proteins. Ochoa's enzyme was used to join up individual uracil molecules – one of the RNA bases – to make polyuracil molecules. This 'poly-U' was then added to Nirenberg's system and protein was formed consisting entirely of units of the amino acid phenylalanine. This protein, called polyphenylalanine, had been made despite the fact that all the other amino acids were present in the mixture and might just as easily have been used instead. Yet, in the same way, polyadenine RNA and polycytosine RNA were found to make proteins consisting entirely of lysine units and proline units respectively.

It was then discovered that mixed RNA polymers consisting of two or more different bases acted as templates for the synthesis of proteins consisting of mixtures of amino acids. For example, a mixed synthetic poly-adenine-cytosine RNA made protein containing lysine and proline, as expected; but the protein also contained histidine, asparagine and glutamine. So it appeared that these last three amino acids must be coded for by some mixture of adenine and cytosine bases, since they are not made when pure polyadenine or pure polycytosine RNA is used. Similar methods enabled very many of the code words to be worked out. There was also at this time other evidence coming from genetic studies, suggesting that the code was a 'triplet' code – that is, that each amino acid is coded for by a combination of three bases called a triplet. There are $4^3 = 64$ different three-letter code words which can be written down using four different letters. However, there are only 20 amino acids, and this suggested that some of the amino acids must have more than one of the 64 possible code words to represent them.

Nirenberg's method was not accurate enough to work out all the words in the code; and so a more elaborate approach was adopted by Gobind Khorana, an Indian scientist, trained in England, who worked on solving the code in laboratories in Canada and the United States. The method he used was to synthesize

UUU	Phe	UCU	Ser	UAU	Tyr	UGU	Cys
UUC	Phe	UCC	Ser	UAC	Tyr	UGC	Cys
UUA	Leu	UCA	Ser	UAA	stop	UGA	stop
UUG	Leu	UCG	Ser	UAG	stop	UGG	Trp
CUU	Leu	CCU	Pro	CAU	His	CGU	Arg
CUC	Leu	CCC	Pro	CAC	His	CGC	Arg
CUA	Leu	CCA	Pro	CAA	Gln	CGA	Arg
CUG	Leu	CCG	Pro	CAG	Gln	CGG	Arg
AUU	Ile	ACU	Thr	AAU	Asn	AGU	Ser
AUC	Ile	ACC	Thr	AAC	Asn	AGC	Ser
AUA	Ile	ACA	Thr	AAA	Lys	AGA	Arg
AUG	Met	ACG	Thr	AAG	Lys	AGG	Arg
GUU	Val	GCU	Ala	GAU	Asp	GGU	Gly
GUC	Val	GCC	Ala	GAC	Asp	GGC	Gly
GUA	Val	GCA	Ala	GAA	Glu	GGA	Gly
GUG	Val	GCG	Ala	GAG	Glu	GGG	Gly

U–uracil (thymine)
C–cytosine
A–adenine
G–guanine
Ala–alanine
Arg–arginine
Asn–asparagine
Asp–aspartic acid
Cys–cysteine
Gln–glutamine
Glu–glutamic acid
Gly–glycine
His–histidine
Ile–isoleucine
Leu–leucine
Lys–lysine
Met–methionine
Phe–phenylalanine
Pro–proline
Ser–serine
Thr–threonine
Trp–tryptophan
Tyr–tyrosine
Val–valine

The genetic code. Of the 64 possible 'triplets', 61 code for the 20 amino acids. The remaining three – the 'stops' – are used to signify the end of a protein chain

short three-base RNA polymers, each equivalent to one code word. Although these code words were not large enough to allow any actual protein synthesis, they were all that was needed to attach the transfer RNA, together with its amino acid, on to a ribosome. Using this technique, Khorana was able to test out all the possible three-letter code words. He thus confirmed the conclusions of Nirenberg and was able to deduce the translation of code words which could not be deduced by Nirenberg's method.

The genetic code is not a random assignment of code words to amino acids. Most amino acids have two or more different code words sharing a basic similarity. For example, all the code words starting with the bases CU code for the amino acid leucine; in this case, the final letter of the code can be any of the four bases. In most cases the final base is indeterminate and can be either uracil or cytosine, or adenine or guanine. This systematic property of the code is exploited by the organism, since a single species of transfer RNA has been found for the amino acid alanine which will recognize all three triplets GCU, GCC and GCA. Crick has suggested that there may be a wobble in the pairing of the third base which in most cases makes the first two letters of the code the crucial ones.

The complete genetic code seems almost too neat and tidy to be true, more reminiscent of the creation of a mathematician than of the 'blind' forces of nature. There can, however, be no possible doubt that the code is correct – it has been experimentally confirmed in living organisms in a number of ways, as well as having been demonstrated in the test tube.

THE EVOLUTION OF MICRO-ORGANISMS 7

The air, the soil, and the water which surround us are full of micro-organisms, some living on the energy they get from the sun, others preying upon one another and on higher organisms. This world of micro-organisms, which holds many clues to the evolution of multicellular creatures, existed unknown to man until the invention of the microscope in the seventeenth century. The first microscope consisted simply of a single lens mounted in a simple frame, with an arrangement for holding the specimen at the best distance from the lens. This type of instrument was exploited by Anthony van Leeuwenhoek, a Dutch cloth merchant, who not only made his own microscopes, but also used them to examine everything from pond water to the scum on his teeth.

He communicated his discoveries by letter to the Royal Society in London. His writing reveals acute observation, although he develops his story with the charmingly irrelevant circumstantial detail of the best gossip. In the following extract he describes how he discovered 'little animals' for the first time – the animals in question being the protozoa now known as *Vorticella*.

The pioneer microscopist Antony van Leeuwenhoek (1632–1723). In his hand he holds one of the many single-lens microscopes he used in investigating the world of micro-organisms

> In the year 1675, about halfway through September (being busy with studying air, when I had compressed it by means of water), I discovered living creatures in rain, which had stood but a few days in a new tub, that was painted blue within. This observation provoked me to investigate this water more

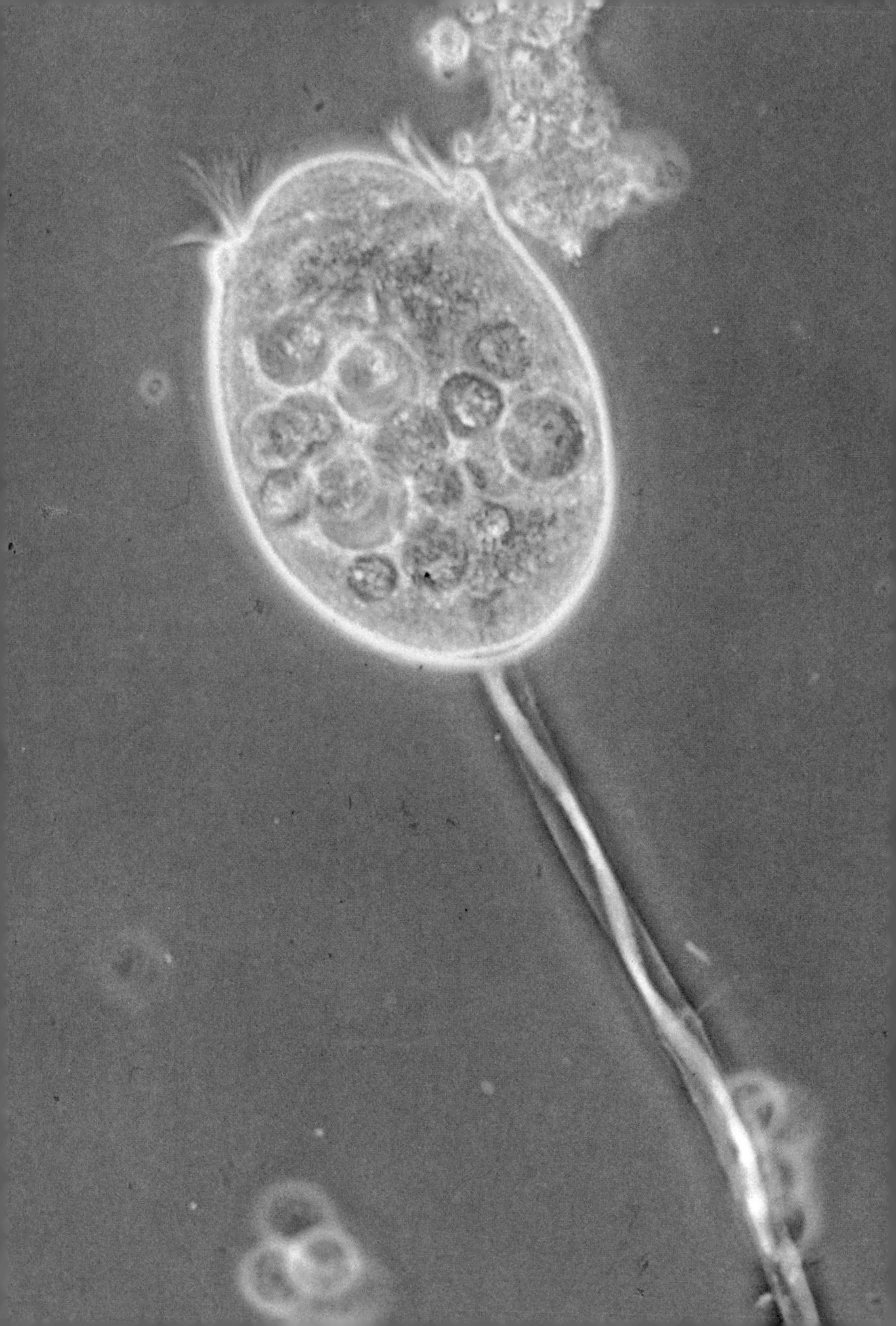

narrowly; and especially because these little creatures were, to my eye, more than ten thousand times smaller than the animalcule which Swammerdam has portrayed, and called by the name of water flea, or water louse, which you can see alive and moving in water with the naked eye.

Of the first sort that I discovered in the said water, I saw, after divers observations, that the bodies consisted of 5, 6, 7 or 8 very clear globules, but without being able to discern any membrane or skin that held these globules together, or in which they were enclosed. When these animalcules bestirred themselves, they sometimes stuck out two little horns, which were continually moved, after the fashion of horses' ears. The part between these little horns was flat, their body else being roundish, save only that it ran to a point at the hind end; at which pointed end it had a tail, near four times as long as the whole body, and looking as thick, when viewed through my microscope, as a spider's web. At the end of this tail there was a pellet, of the bigness of one of the globules of the body; and this tail I could not perceive to be used by them for their movements in very clear water. These little animals were the most wretched that I have ever seen; for when, with the pellet, they did but hit on any particles or little filaments (of which there are many in water, especially if it hath stood some days), they stuck entangled in them; and then pulled their body out into an oval, and did struggle, by strongly stretching themselves, to get their tail loose; whereby their whole body sprang back towards the pellet of the tail, and their tails then coiled up serpent wise, after the fashion of a copper or iron wire that, having been wound close about a round stick, and then taken off, kept all its windings. This motion, of stretching out and pulling the tail, continued; and I have seen several hundred animalcules, caught fast by one another in a few filaments, lying within the compass of a coarse grain of sand. (From C. Dobell, *Antony van Leeuwenhoek and his Little Animals.)*

Opposite: Vorticella – the unicellular micro-organism described by van Leeuwenhoek in his letter to the Royal Society quoted on this page. Below: photograph of a microscope made by van Leeuwenhoek

Van Leeuwenhoek went on to discover and describe many more of these little protozoan animals and was probably the first man actually to see the cells of bacteria. It could justifiably be said that he discovered a third world – the world of micro-organisms.

This world of micro-organisms includes the single-celled algae, or plant-like micro-organisms, as well as the single-celled animals, or protozoa, although no hard and fast distinction between animals and plants is really possible when it comes to these single-celled creatures. Micro-organisms can be divided into two quite distinct groups: the eucaryotes and the procaryotes. Protozoa, algae and fungi as well as higher animals and plants are eucaryotes, characterized by the possession of a nucleus – with chromosomes covered in a coat of histone – surrounded by a nuclear membrane. Bacteria and some closely related groups are procaryotes, which are characterized by a circular chromosome of naked DNA not surrounded by a nuclear membrane. The distinction between eucaryotic and procaryotic organisms is a very important and fundamental one, so that from the evolutionary point of view eucaryotic micro-organisms are much closer to higher plants and animals than are procaryotic micro-organisms.

Micro-organisms have evolved ways of living in an incredible variety of environments. By looking at some of the major groups of micro-organisms we can see what evolutionary selection has achieved at the level of the cell. It is also possible to trace some of the pathways by which micro-organisms have evolved, although there are still large gaps in our knowledge.

Protozoa – the microscopic animals

There are four major classes of protozoa: the ciliates; the amoeboid protozoa; the spore-forming protozoa; and the flagellates, which can also be classified as algae.

Vorticella, the protozoan described by van Leeuwenhoek, is a ciliate. This name is given to a group of micro-organisms which possess tiny moving hairs called cilia, and have two nuclei – a macronucleus and a micronucleus. In *Vorticella* the cilia are arranged in a ring round the mouth and beat in such a way that food, consisting mostly of bacteria in the surrounding water, is swept into its gullet. *Vorticella* is attached to a solid object by means of a stalk, but if food becomes short it can cast off its stalk and swim in search of more favourable feeding grounds.

Paramecium, another very well-known ciliate, is covered all over with short cilia which beat in unison, so pushing it through the water. The action of these cilia is believed to be co-ordinated by a network of fibres called the neuromotor apparatus. Bacteria are swept into its gullet as it swims, forming food vacuoles which circulate through the cytoplasm while digestion proceeds. When digestion is completed, the waste contents of the vacuole are discharged through a specialized part of the cell wall – the cell anus. *Paramecium* and many other protozoa have a special organelle – the contractile vacuole – which regulates the amount of water present in the cell. In fresh water environments a great deal of water enters the protozoan cell by osmosis and this must be actively excreted by the cell, which would otherwise burst. The contractile vacuole of *Paramecium* can be seen doing this under the microscope, moving rhythmically like a pump, at a speed which depends on the salt content of the water.

Paramecium has a macronucleus and a micronucleus. The micronucleus stores genetic information, and, as with nuclei of other cells, can be involved in sexual or asexual division processes. The macronucleus, on the other hand, contains very much more DNA than the micronucleus, and can be thought of as a means of amplifying the latter's genetic information. During mating the macronucleus is destroyed and then remade from the micronucleus. The macronucleus is concerned with the ordinary business of running the cell and the micronucleus with passing on genetic information from one generation to the next. By this means *Paramecium* has evolved within one cell two separate functions which, in multicellular organisms, are performed by two separate tissues. The micronucleus of *Paramecium* is comparable to the male or female reproductive organs (germplasm) of higher organisms and the macronucleus is comparable to the rest of the body (somatoplasm). This specialization within the cell is also shown in the contractile vacuole, neuromotor apparatus, gullet and anus of *Paramecium,* which might be compared, according to the functions they serve, to the kidney, nervous system and digestive tract of multi-cellular organisms.

There are many other types of ciliates; some of them have their cilia fused together into little triangular plate-like structures called membranelles, such as those around the mouth of *Stentor*. Some ciliates are carnivorous: *Didinium,* for example, feeds on *Paramecium,* piercing it with a special sharp mouthpiece and then sucking out its insides.

Below: a colony of the alga Pediastrum, grouped in spiral formation. Opposite top: trumpet-shaped Stentor. A comparative giant among protozoa, it may exceed 2 mm in length and is capable of engulfing more than a hundred euglenas (see page 129) a minute. Opposite bottom: a paramecium

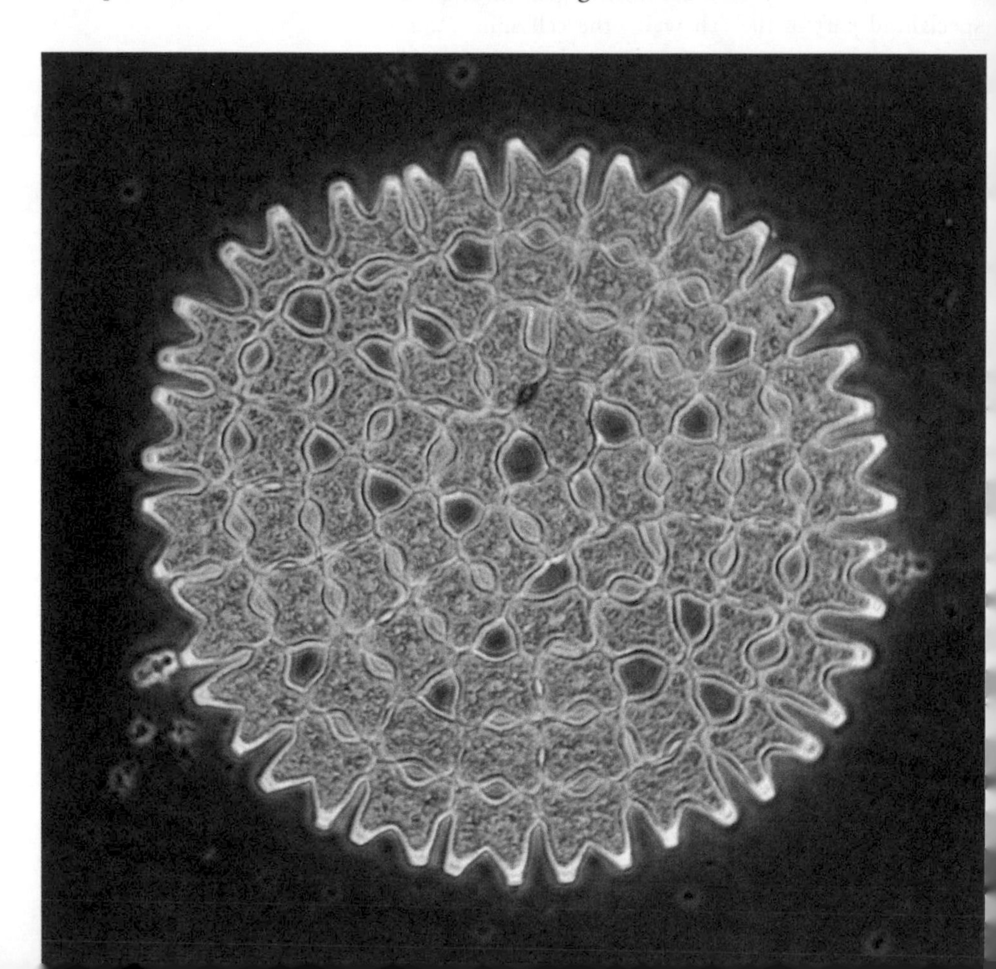

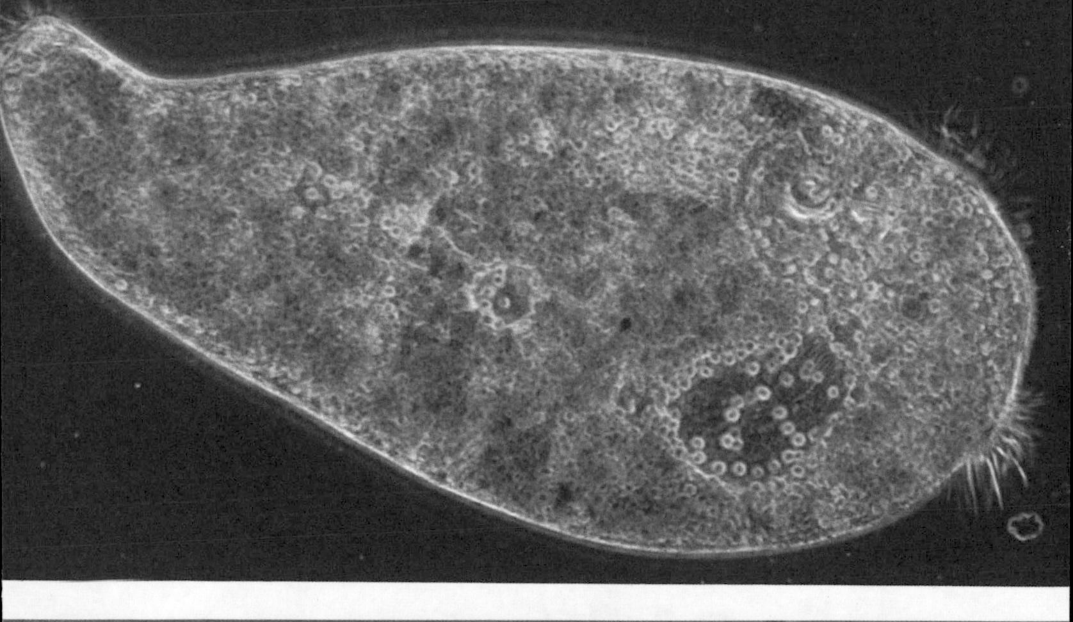
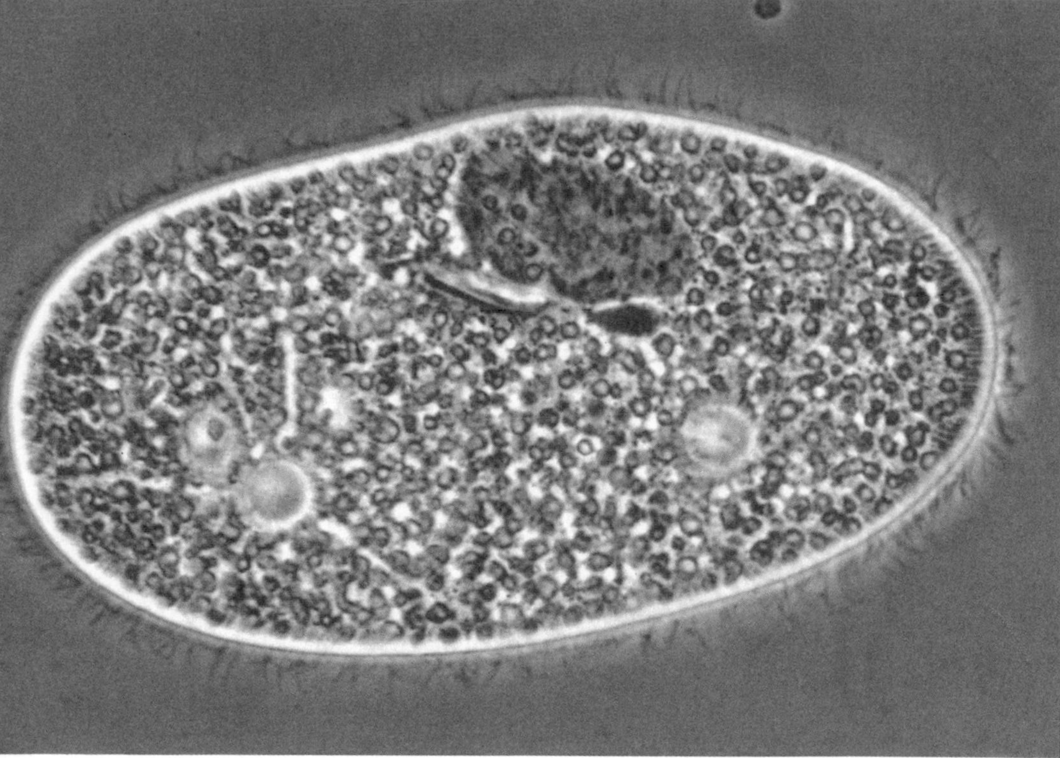

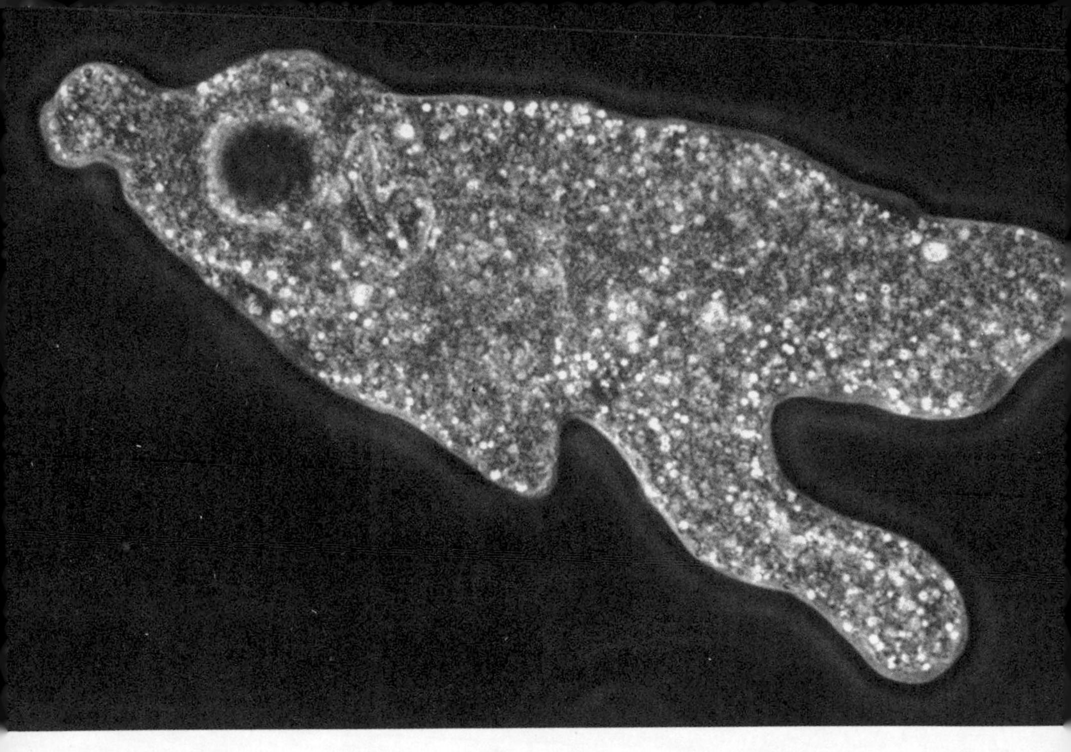

Above: a common amoeba of a kind that lives harmlessly in the human intestine, feeding on particles – including other parasites – in the intestinal contents. Opposite: some of the nineteenth-century naturalist Ernst Haeckel's exquisitely detailed drawings of silica skeletons secreted by radiolaria – an amoeboid protozoon

There was a time when a book about the cell could not be written without first describing *Amoeba,* which was thought of as an ideal cell. There is of course no typical or ideal cell, so the amoeba no more deserves this special treatment than other cells. Amoebae, do, however, show an uncanny resemblance to certain animal cells, such as the white blood cells; and the study of amoebae has provided some basic understanding of cell movement and of feeding by the engulfing of particles. Amoebae move by stretching out limb-like projections of their cytoplasm – the pseudopodia – and then 'flowing' into them. The pseudopodia are also used to surround food particles and take them inside the cell. The food particles are then contained in a digestive vacuole into which enzymes are released to break down the food for use by the cell. There are many species of amoebae and they are found in a wide range of environments both as free-living organisms and as parasites in the mouth and gut of animals and man. Some species have a radically different appearance, but are nevertheless classified with the amoebae

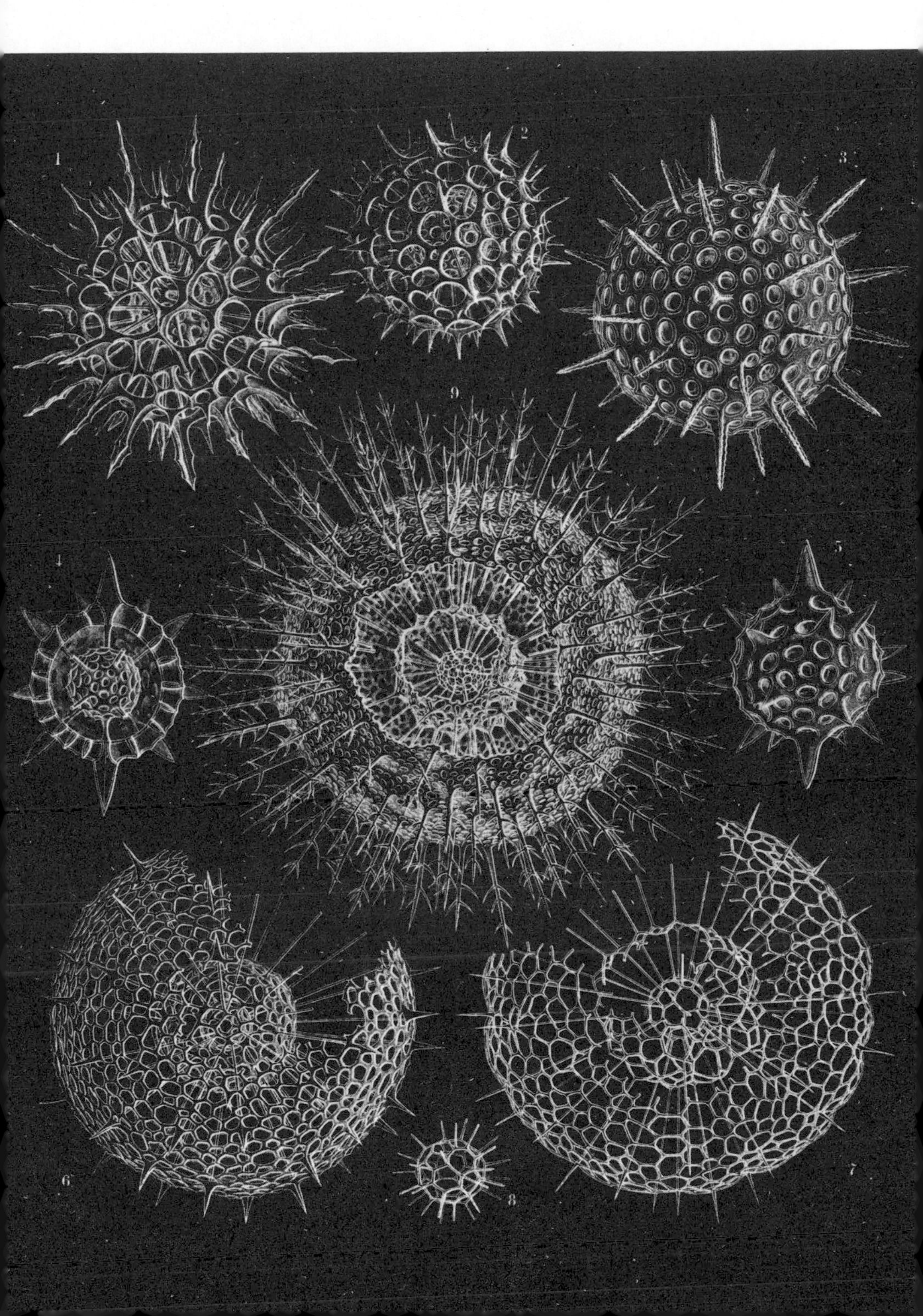

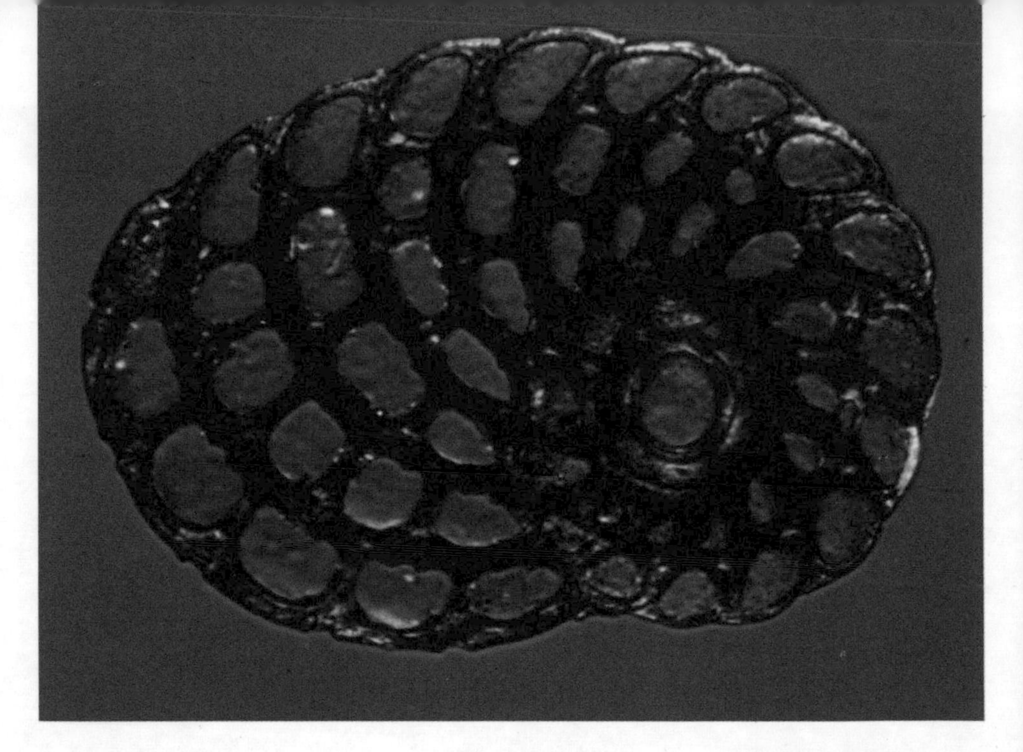

Above: about the size of a pin-head, the chalky shell of a foraminiferan amoeba. In life, slender pseudopodia would have projected through the many tiny pores in the shell's surface. Opposite: a photomicrograph of radiolaria enlarged 180 times

because they form the characteristic pseudopodia. The foraminiferan amoebae have a cell skeleton made of chalk (calcium carbonate) through which many thin pseudopodia can poke out. These animals are remarkable for the size they attain – up to several millimetres in diameter – which lends some support to the view that protozoa are non-cellular rather than unicellular. The radiolarian and many heliozoan amoebae have cell skeletons made of silica. Some of these are reminiscent of the sculptures of Barbara Hepworth inspired by her studies of geometrical construction. D'Arcy Wentworth Thompson, in his classic book *On Growth and Form*, compared the skeletons of radiolarians to the structure of snow crystals and speculated about the physical forces controlling their formation. These studies suggested to him that cells, parts of cells and tissues may yet reveal in their structure that they were originally constructed through physical processes akin to crystallization.

Spore-forming protozoa – such as the malaria parasite – are all parasitic. They have a stage in their life

cycle during which they reproduce by forming spores produced by rapid nuclear division. This ensures that the parasite can achieve large numbers in the blood of the host at certain times in the life cycle. The malaria parasite of man and animals is transmitted from one animal to another by the bite of a mosquito. In man malaria typically causes periodic fevers occurring at night at intervals of a few days, and these fevers are caused by the massive increase in numbers of the malaria parasite by spore formation. Mosquitos only bite at night and it is believed that the increase in numbers of the malaria parasite at night is an adaptation to this, giving the parasite the best chance of being sucked up by the mosquito, and so of reaching new hosts.

Plants and plant-animals

Some flagellates are plant-like and some are animal-like; some can be classified with the protozoa and some with the algae. They have a fairly definite shape and a flagellum at one end which propels them through the water with a whiplash movement. Plant-like flagellates such as *Euglena* possess chlorophyll granules, but can wriggle along by moving their bodies. This euglenoid movement is regarded as an animal-like property. *Euglena* also has an eyespot – a light red granule which is believed to be responsible for its tendency always to move towards light.

Volvox is a spherical organism a few millimetres in diameter made up of large numbers of individual cells (up to 15,000) arranged side by side as if on a spherical surface. Each one of these cells is similar to an individual *Euglena* cell. The individual cells of *Volvox* are connected by protoplasmic threads which probably serve as a means by which messages are transmitted – so ensuring that the flagella all beat in a co-ordinated way. The cells also show some specialization: those cells at the front of the organism have larger eyespots than those at the back, whereas the cells at the back are capable of reproduction and those at the front are not. These features show that *Volvox* is not just a colony of individual euglenoid animals, but rather a primitive

Opposite: a euglena in which the pattern of contractile fibres which permit contraction and elongation of the body is highly conspicuous (see also page 129)

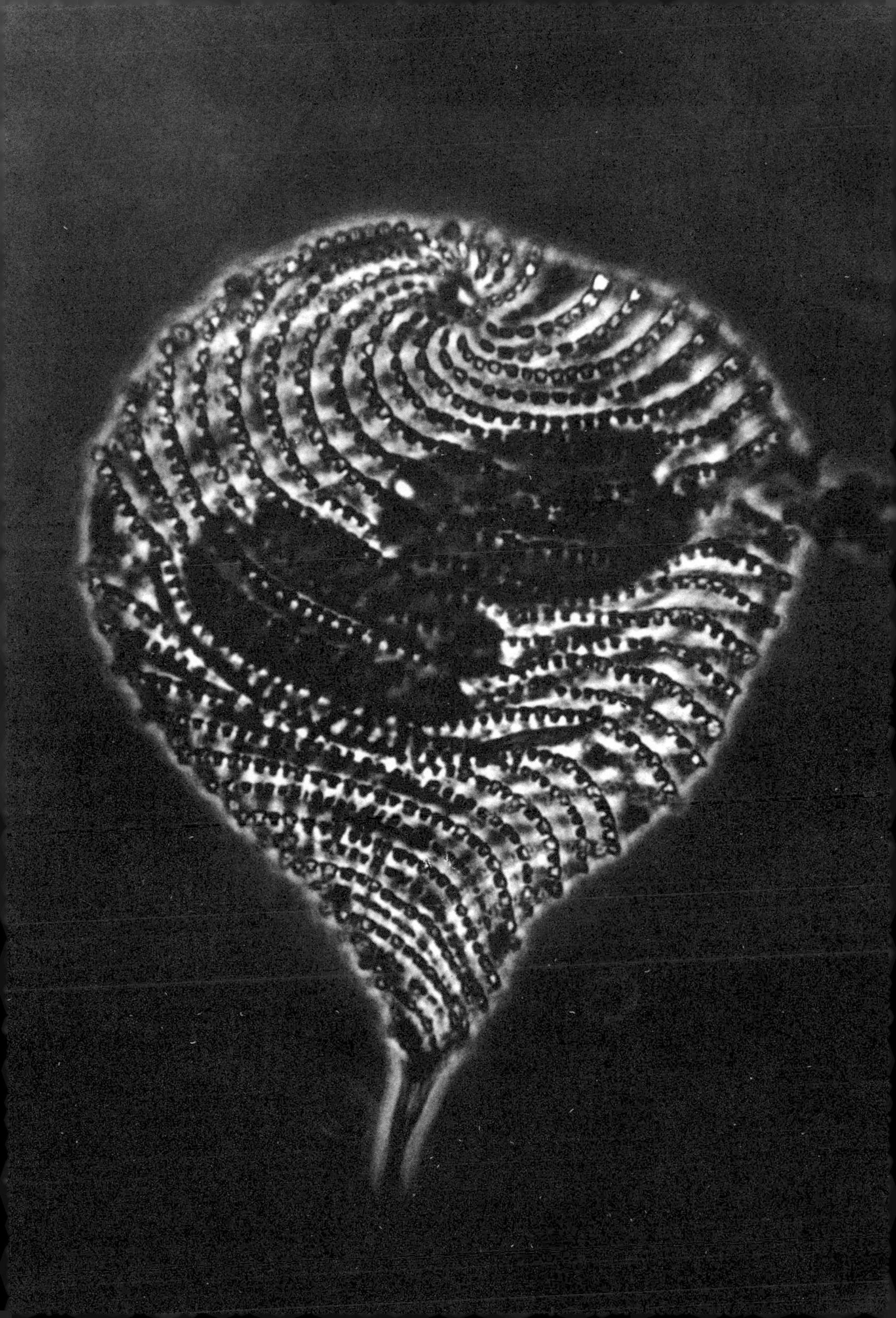

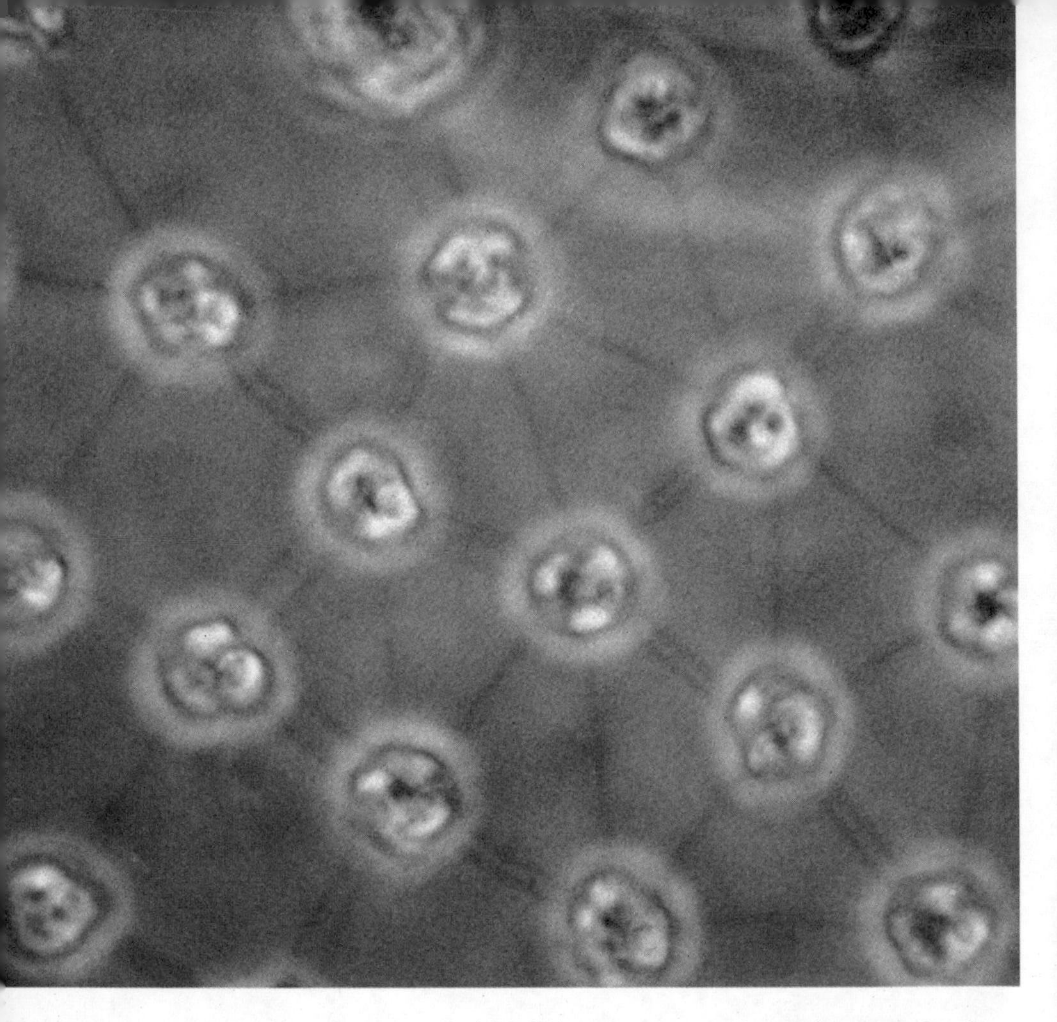

Above: a small area of the surface of a spherical colony of Volvox, 1,750 times enlarged. Opposite: photomicrograph of a euglena, showing nucleus, red eyespot and green chlorophyll granules, about 2,000 times life size

multi-cellular organism. In fact *Volvox* has often been compared to the embryonic blastula, found as a stage in the early development of vertebrates. Both *Volvox* and the blastula are simply a sphere of cells. It is not difficult to imagine that multicellular organisms evolved from single-celled organisms by going through a series of colonial forms resembling *Volvox*.

There are other flagellates which are much more animal-like than *Euglena* and do not possess chlorophyll. Some of these are free-living, such as the marine dinoflagellates, which feed on bacteria. One of these, *Noctiluca*, gives out light and can be seen as sparks of luminescence in the sea when it is disturbed at night.

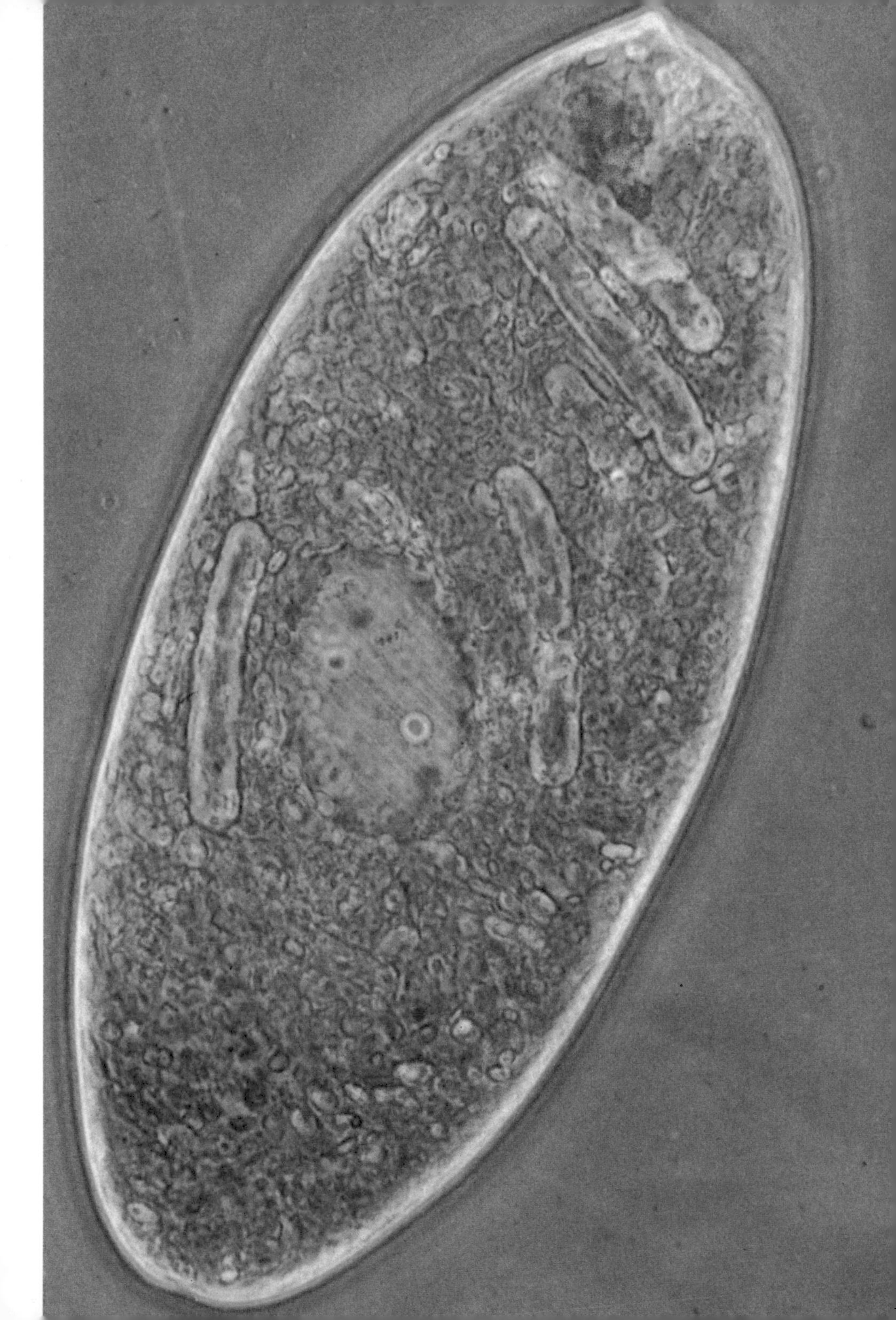

There are also parasitic flagellates, such as the trypanosome, which causes sleeping sickness and is spread from one host to another by the bite of the tsetse fly. One curious flagellate, *Trichonympha*, living in the gut of termites, is able to digest the pieces of wood that are swallowed. The termites themselves are unable to digest wood and so, without *Trichonympha*, would not only get the most fearful indigestion, but would soon starve. This type of association of two animals which is of mutual benefit to both is called symbiosis; and the parasitic type of relationship can often evolve into a symbiotic one. *Trichonympha* was probably merely a gut parasite of the termite before the termite came to depend on a woody diet.

There are some flagellates which exist as two distinct species that are identical in appearance, except that one possesses chlorophyll, while the other does not. *Euglena* itself can be made to lose its chlorophyll – by growing it at high temperature or by exposing it to ultraviolet light – and will carry on living quite happily, provided there is plenty of food (sugars) in solution in the immediate environment. And once the chloroplasts have been lost they cannot be regained, because they are separate hereditary particles with their own DNA. This situation is common among the dinoflagellates, which often exist as both green photosynthetic forms and colourless leucophytic forms. However, some dinoflagellates have evolved so far away from the independent photosynthetic forms as to live parasitically in marine animals.

Protozoa and algae almost certainly evolved from some common eucaryotic ancestor which contained one type of particle – the forerunner of both the mitochondrion and the chloroplast. For the DNA in both the mitochondrion and the chloroplast has no histone protein associated with it and is not surrounded by a membrane. In these and other respects (discussed in chapter 2) the chloroplast and the mitochondrion resemble procaryotic cells, and indeed it seems most probable that eucaryotic cells have evolved from the symbiotic association of a primitive eucaryotic cell

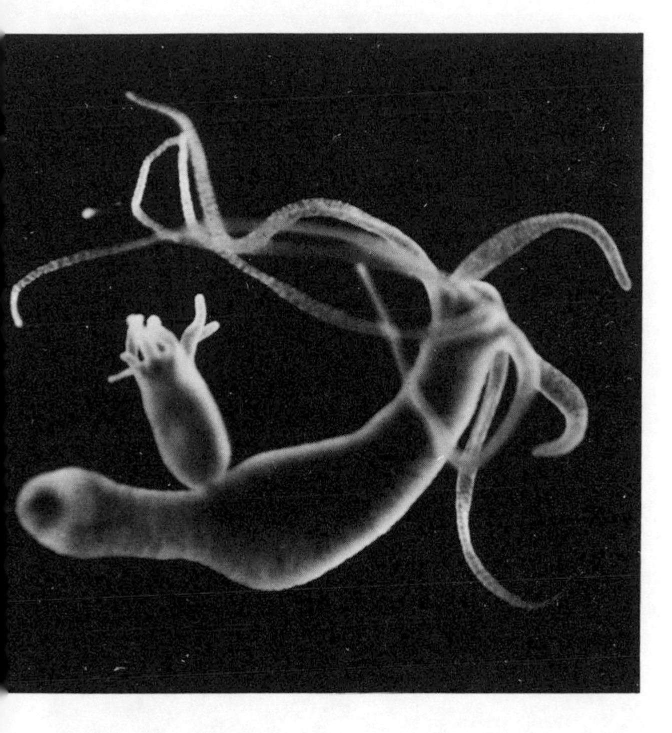

Hydra – a multicellular animal which, by associating with unicellular algae in a symbiotic relationship, has retained the plant-like ability to derive energy from sunlight

with a procaryotic cell. This symbiotic procaryotic cell must then have evolved within the eucaryotic cell into separate mitochondria and chloroplasts. Plant cells must have evolved both types of particle, whereas animal cells might have evolved without ever having separate chloroplasts. Alternatively, animal organisms could have (at least among the flagellates) gone on evolving from plants by the simple loss of chloroplasts. The higher plants as we know them today, however, have become so specialized that they could not live in nature without their chloroplasts.

It is interesting that some protozoa and fungi – and even multicellular animals such as hydra, the fresh-water anemone – have found it advantageous to regain the ability to get energy from the sun. They have achieved this by associating in a symbiotic relationship with unicellular algae. Some species of *Paramecium* contain hundreds of green algal particles. Lichens, the unprepossessing organisms that grow on rocks and

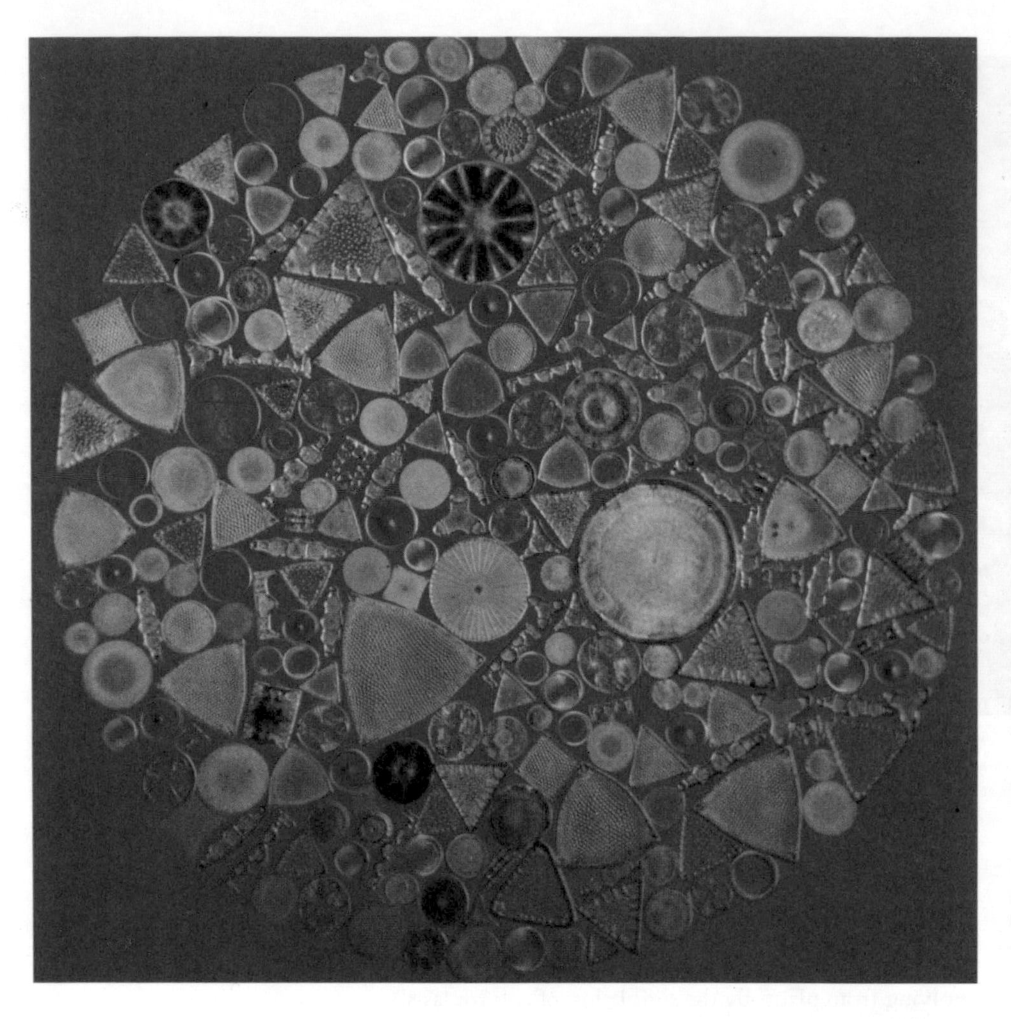

An artificially arranged selection of diatoms, magnified 95 times. Transparent single-celled algae, diatoms possess cell walls impregnated with silica and acquire their rainbow colours by refracting light

the bark of trees, are formed by the symbiotic association of a fungus and an alga. It might be said that evolution, like history, repeats itself, although in fact nothing ever happens in exactly the same way twice.

Besides the photosynthetic flagellates already mentioned, there are many other kinds of algae – including the diatoms, the common red and brown seaweeds and several other less well known groups. It seems most likely that these have all evolved from a common type closest to the flagellates. The fungi are traditionally considered to be plants and are studied by botanists, but

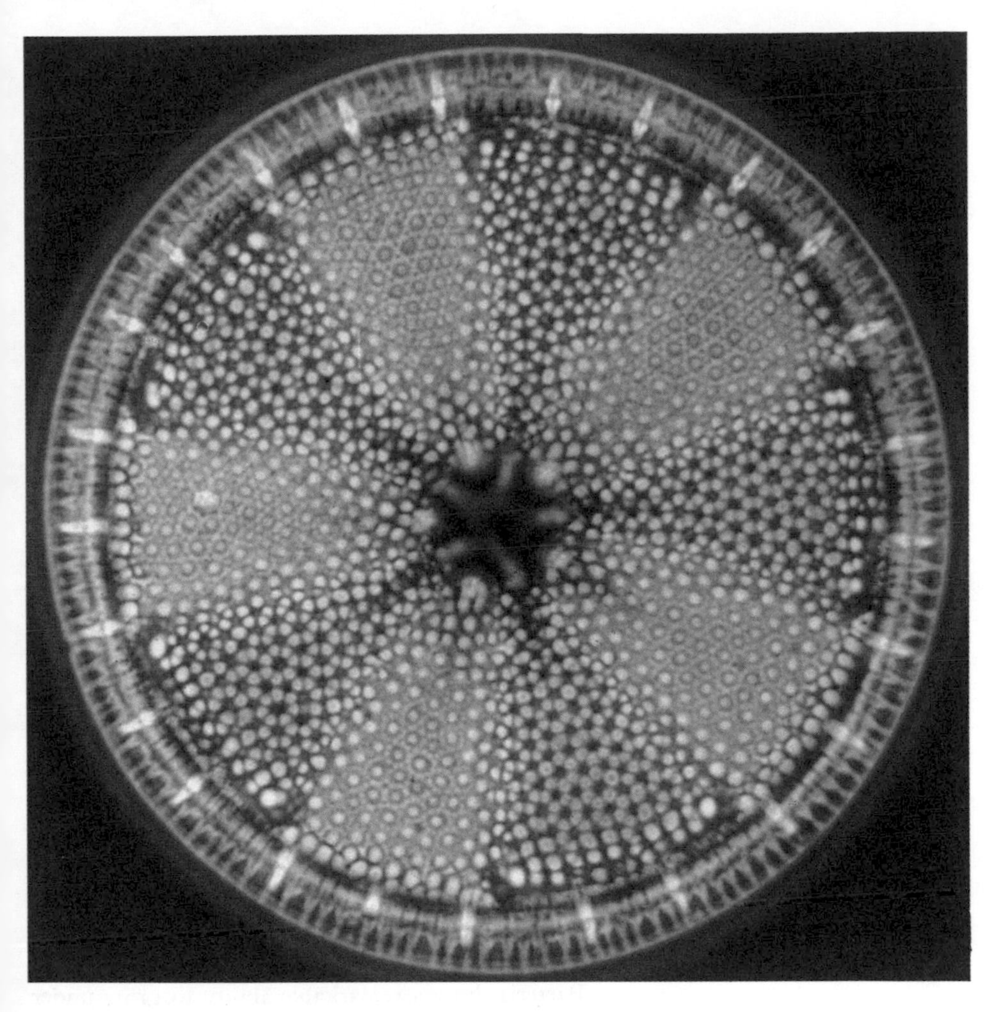

A single diatom, magnified 600 times

they are really more closely related to the protozoa than to the algae. The fungi have no chloroplasts and so are dependent upon organic material, often refuse, for their energy. In this way they resemble the protozoa – although they cannot engulf particles of food as protozoa can. Instead, they rely upon breaking down their food with digestive enzymes which they excrete, and then absorbing the food in solution. The tough chitinous cell wall characteristic of fungi is also similar to the cell walls of many protozoa and quite distinct from the cellulose wall of the algae.

Bacteria

The largest procaryotes – bacteria and related organisms – are about the same size as the smallest eucaryotic cells, and can easily be seen under a good light microscope. Different shapes of bacteria – rods, spheres, spirals and strings – can be seen. And some kinds of bacteria are equipped with flagella – which differ in structure from the protozoan flagella. However, what distinguishes the bacteria and other procaryotes from the eucaryotes is not size or shape, but the radically different organization of the nucleus with its naked DNA, and the lack of intracellular organelles such as mitochondria and chloroplasts.

The bacterial cell contains a structure called the mesosome, which is formed from the cell membrane. The DNA of the cell seems to be attached to the mesosome, which is believed to play an important role in the processes of cell division and cell-wall synthesis. The mesosome also contains enzymes, associated with electron transport, which are equivalent to those found in the mitochondria of eucaryotic cells. Most bacteria are surrounded by a rigid cell wall quite different from the cellulose cell wall of plants. The bacterial cell wall contains an unusual sugar (muramic acid), and an unusual amino acid (diaminopimelic acid) which is not found in proteins. These compounds are linked together with other amino acids to form an elaborate layered meshwork, whose detailed molecular structure is just beginning to be understood.

Bacteria show a remarkable ability to grow under almost every conceivable condition. There are bacteria which grow in hot springs at a temperature as high as 92°C – which would kill almost any other living thing. And there are bacteria which actually feed on petroleum and grow in fuel tanks, causing a great deal of trouble to man. Most bacteria, however, feed on existing organic materials – dead animals and plants – causing putrefaction and decay; but there are also some bacteria which get their energy from the sun by photosynthesis.

Bacterial photosynthesis differs somewhat from

plant photosynthesis. In plant photosynthesis hydrogen is removed from water and its energy is boosted; oxygen is produced as a result. However, in bacterial photosynthesis hydrogen is taken from hydrogen sulphide or organic compounds. Carbon dioxide gas is often used as the source of carbon and, in the green sulphur bacteria, reacts with hydrogen sulphide gas, giving a sugar, water and sulphur. This reaction is analogous to that for plant photosynthesis, where water rather than hydrogen sulphide acts as the hydrogen donor. These photosynthetic reactions are conducted in a special structure – the chromatophore – which could be compared to the chloroplast.

Photosynthetic bacteria can only grow in anaerobic conditions and so are found in the bottom of stagnant ponds. Aerobic algae grow on the surface, where there is oxygen, and the photosynthetic bacteria grow beneath them. The bacteria utilize whatever light in the red and infrared parts of the spectrum remains unabsorbed by the algae.

There are also bacteria which live in soil by getting energy from the oxidation of inorganic materials such as minerals, rather than from the oxidation of organic materials such as sugars. Other bacteria get energy by oxidizing or 'burning' hydrogen gas. The sulphur bacteria oxidize hydrogen sulphide to inorganic sulphur and then to sulphuric acid. The iron bacteria obtain their energy by oxidizing ferrous iron to ferric iron and the nitrifying bacteria obtain their energy by oxidizing ammonia or nitrous acid.

Some bacteria are able to form inactive resting spores, which enable them to survive lengthy periods of food shortage, and which are highly resistant to heat and attack by chemical agents. *Clostridium* bacteria form spores which may find their way into tinned food and there germinate, forming the highly poisonous botulinus toxin. This substance is a protein which has a very high affinity for the cells of the nervous system. One milligramme is enough to kill a million guinea pigs. It is the most poisonous substance known, yet scientists still do not understand exactly how it acts.

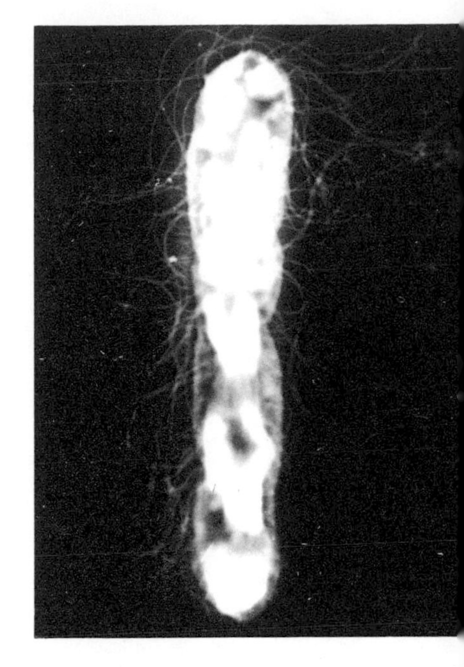

The bacterium Bacillus proteus (above), and (below) a colony of Clostridium – the bacteria producing the deadly botulinus toxin

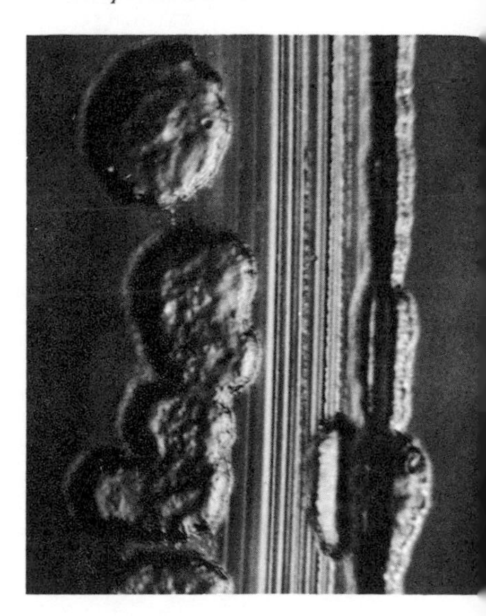

There has been speculation in the past that some groups of bacteria are perhaps closer to the eucaryotes than to other bacteria. The blue-green algae (myxobacteria) were thought at one time to have closer affinities to algae than to bacteria. These organisms are usually photosynthetic and have, moreover, the same kind of aerobic photosynthesis as plants do. They can glide along across a solid surface – although it is not known how – and they have cell walls similar to those of bacteria. The blue-green algae can 'fix' sufficient nitrogen from the atmosphere to make all their own amino acids. This makes them very independent and able to grow in very bleak places. In fact, it seems possible that these organisms were among the first in the course of evolution to conquer the land. Support for this belief comes from the observation that blue-green algae were among the first organisms to be found repopulating the island of Krakatoa after it exploded volcanically in 1883.

Another group of bacteria called the *actinomycetes* are filamentous and at one time were thought to be related to the fungi; but they show the typical procaryotic nuclear organization. This group includes the tubercle bacillus, which causes tuberculosis, and a closely related bacterium which causes leprosy. *Streptomyces*, another type of actinomycete, is rather like a miniature fungus, and has been found to secrete several useful antibiotics, including the one named after it – streptomycin. The close superficial similarity of *Streptomyces* to a miniature fungus is an example of convergent evolution. The same type of structure has evolved more than once, because it has independently proved advantageous to both *Streptomyces* and the fungi.

Syphilis and several other human and animal diseases are caused by the spirochaetes, a group of procaryotes distant from the common disease-causing bacteria. Free-living species of spirochaetes also exist, and they can be found in polluted water with a low oxygen content. The body of the spirochaete is a flexible spiral which is, as it were, wrapped round an axial filament.

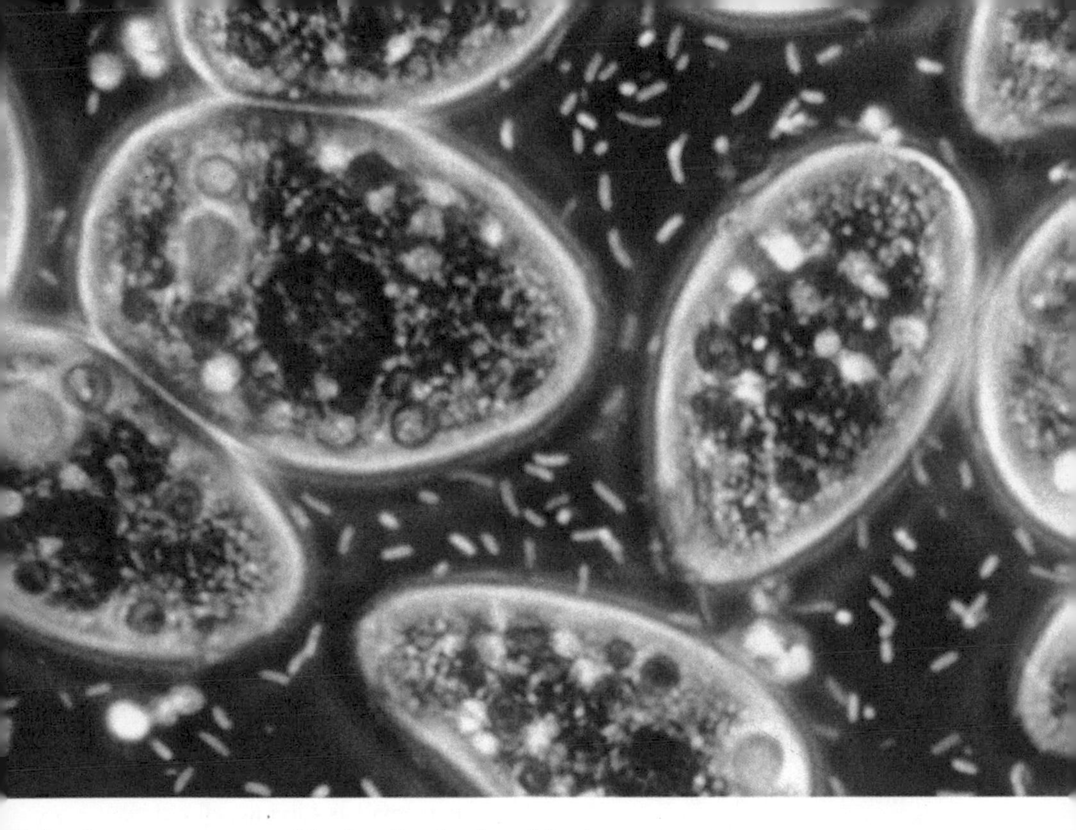

Single-celled Tetrahymena (the relatively large organisms in this photomicrograph) feeding on bacteria (much smaller)

Spirochaetes move around by bending back and forth in snake-like coils.

The smallest living things

The smallest free-living organisms known are an unusual group of procaryotes, which was first recognized by Pasteur as the causative agent of pleuropneumonia in cows. These organisms, found to pass readily through porcelain filters which would not allow the passage of bacteria, were initially confused with viruses. As early as 1898 these organisms were successfully grown in a broth-like medium.

A number of strains of this organism have now been isolated, including one, causing inflammation of the urinary duct in man (urethritis), called the Eaton agent. Because these organisms are so different from bacteria, and are now clearly seen to bear no resemblance to viruses, they have been called *Mycoplasma* or PPLO, which stands for pleuropneumonia-like organisms.

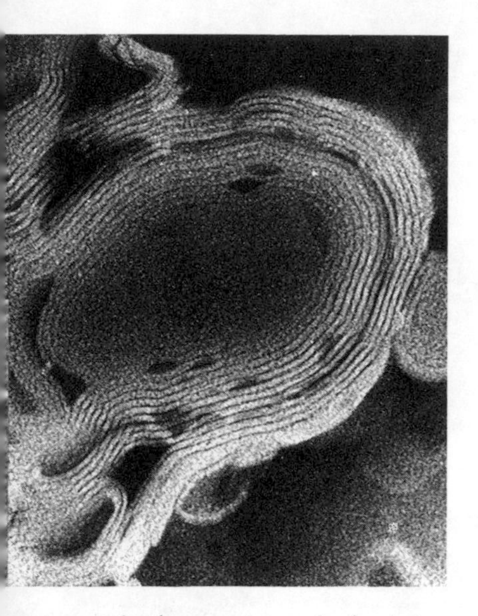

An electron micrograph of a cholesterol/lecithin mixture, magnified some 120,000 times, showing molecular leaflets of fat

Some species of *Mycoplasma* are a little larger than the smallest bacteria, but the smallest *Mycoplasma* are one-tenth the diameter of a typical bacterium. This is smaller than many viruses and only a thousand times the diameter of a hydrogen atom.

Mycoplasma organisms have been found to have a life cycle peculiar to themselves. They start life as 'elementary bodies' which increase in size to form cells which may then break up into small elementary bodies once more. Alternatively, the cells can continue to divide and grow as large cells. *Mycoplasma* are conventional organisms in so far as they consist of DNA, RNA and protein and are surrounded by a membrane which, under an electron microscope, appears the same as the membranes surrounding the cells of other organisms. The membrane does however contain the substance cholesterol, which is present in the cell membranes of higher animals but not in the membranes of bacteria. But despite the very small size of *Mycoplasma* there is no reason to believe that it is some radically new form of life – though there is reason to believe that it is about as small as a living cell can be.

Biochemists calculate that at least one hundred different enzymes are needed to catalyse the reactions essential to maintain *Mycoplasma*. Yet there only seems to be enough DNA present to code for about fifty – far too few, apparently, unless *Mycoplasma* is organized very economically and has some enzymes doing more than one job. The full-grown *Mycoplasma* cell is about a hundred times the minimum volume a cell could occupy and still have one of each kind of enzyme present; whereas the elementary body is only eight times this minimum volume. Smaller sized organisms might in theory be found, but it is doubtful whether a smaller organism would be able to find an environment in which it could compete successfully against other organisms. A smaller organism constructed from the minimum number of components would be living very dangerously. If it possessed only one molecule of each essential enzyme in its cell, the chances of death by accident due to one essential component failing for any

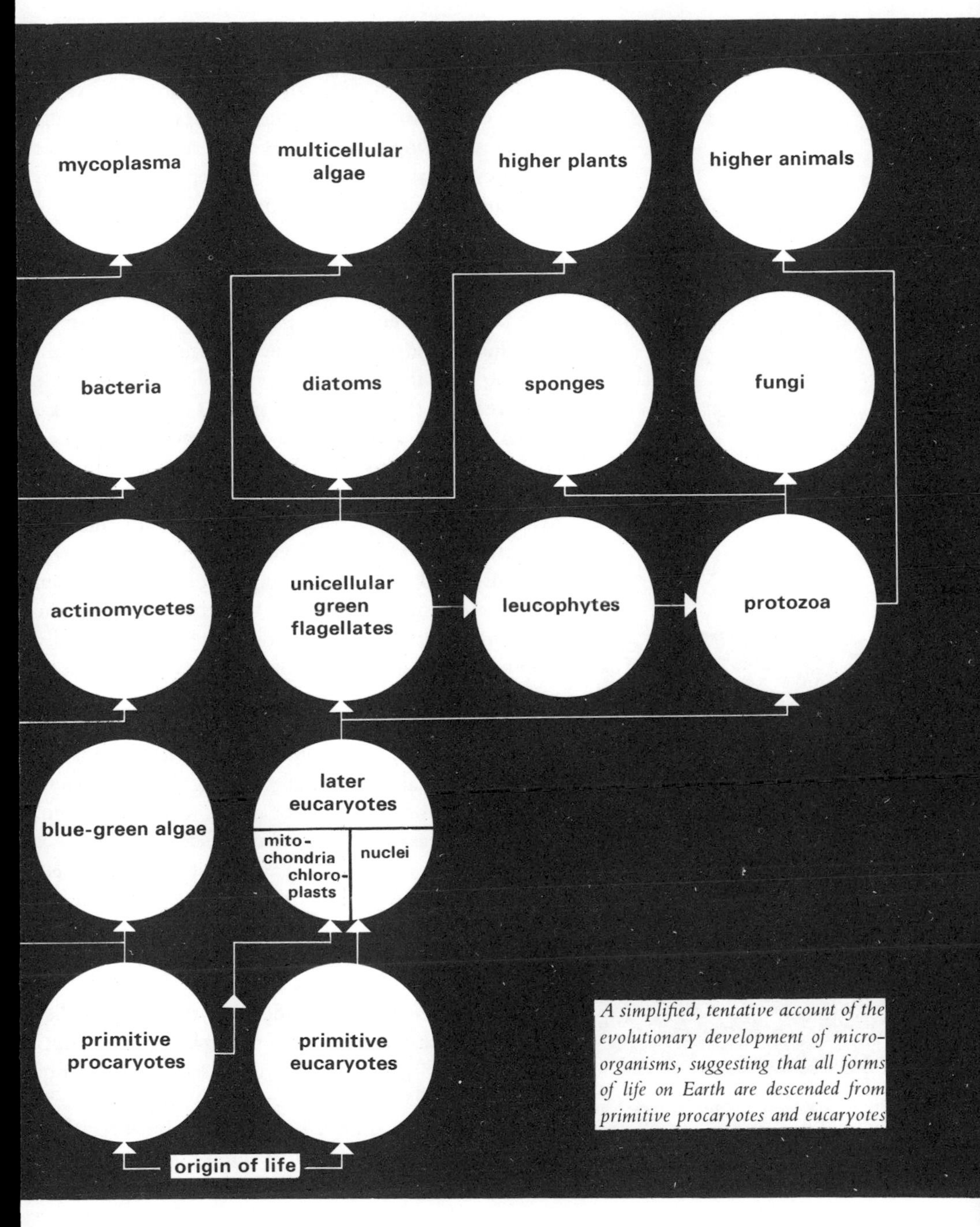

A simplified, tentative account of the evolutionary development of micro-organisms, suggesting that all forms of life on Earth are descended from primitive procaryotes and eucaryotes

reason would be extremely great. For this reason it seems likely that *Mycoplasma* closely approaches the minimal organism. All *Mycoplasma* are parasites of higher organisms, and so, although they are the smallest known living things, there is no reason to believe that they are any closer to the first cellular organisms than are other micro-organisms.

The 'life' of viruses

Viruses first attracted attention as agents of disease, but for a long time remained a puzzle. They were so small that they could not be trapped by filters; nor could they be seen under the light microscope. It was for these reasons that they were given the name virus – which means a sort of living venom or poison. They are 'alive' in the sense that they can increase in numbers by replicating themselves – but then this does not seem so different from the ability of a seed crystal to multiply in a saturated solution. Indeed, the superficial similarity of viruses and crystals was emphasized in 1935 when viruses were crystallized for the first time; but the growth and structure of even the simplest virus is really so complicated that the analogy with crystals cannot be pressed.

In its crystalline form (above) the sowbane mosaic virus seems to exist on the very borderline between the worlds of inanimate and living things

All viruses so far analysed consist of *either* RNA and protein, *or* DNA and protein. The nucleic acid is the genetic material of the virus carrying the instructions that enable more viruses to be made, while the protein forms a protective coat round the nucleic acid. Viruses differ from all other organisms in not containing *both* RNA and DNA. Furthermore, it is only in viruses that RNA is known to act as genetic material and not in the subservient role of carrying genetic messages from nucleus to cytoplasm.

Viruses cannot 'live' outside the cytoplasm of other cells. In the early days of virus research many attempts were made to grow them in the test-tube in various elaborate media, but they would not grow unless cells were present too. Today no one finds this surprising, since viruses are so evidently unlike whole cells in structure: in fact they are much more like pieces of a cell. Viruses vary in size from about one hundred-thousandth to one five-thousandth of a millimetre; and since a molecule of the protein haemoglobin is six millionths of a millimetre in diameter, viruses may sometimes approach molecular dimensions.

All viruses are agents of disease. They are cellular parasites which infect animal, plant and bacterial cells. There are even a few 'viruses of viruses', called satellite viruses, which can only multiply when they infect a bacterium together with what is called a helper virus.

One of the smallest viruses known is a parasite of the colon bacillus. It is called simply MS2. MS2 contains only enough RNA to code for three proteins. The first is the coat protein, the second is an enzyme which is essential for making more virus RNA, and the third, called the 'maturation factor', helps in some way to assemble the virus particles.

Viruses often enter tissues with the assistance of biting, sucking insects, or may be spread about in droplets of moisture or on wind-borne dust particles, coming into contact with cells by accident. When the virus encounters a cell of the right kind (and viruses are very particular about the cells they infect), the virus nucleic acid usually penetrates into the cell leaving the protein coat outside. Some viruses have special enzymes and others elaborate injecting devices to help them get their nucleic acid into a host cell.

Once inside the cell the virus nucleic acid usually instructs the cell to make more virus nucleic acid and protein coats, using the cell's own machinery – the ribosomes and transfer RNA. Several hundred virus particles may be made inside a single cell before the cell is finally destroyed and the particles liberated to infect other cells.

However, some viruses entering the cell do not multiply and destroy; instead they modify the cell and multiply with it. It has been known for some years that the DNA of bacterial viruses can become integrated into the bacterial DNA instead of multiplying and destroying the cell. More recently it has been shown that some DNA viruses can become integrated into the

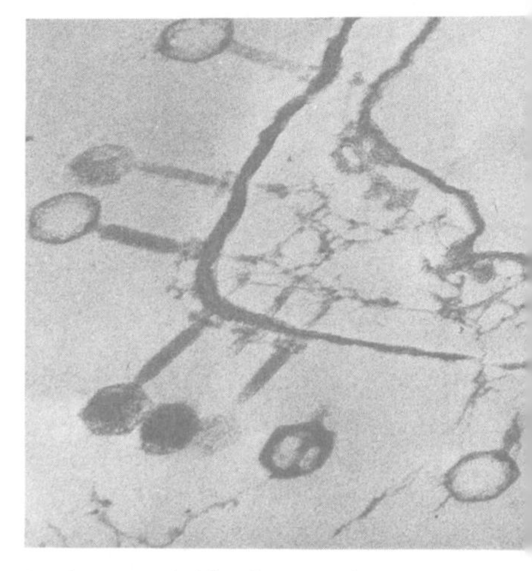

In this remarkable electron micrograph (above), viruses are shown apparently infecting a bacterial cell by injecting their threadlike DNA through the cell wall. Below: an influenza virus

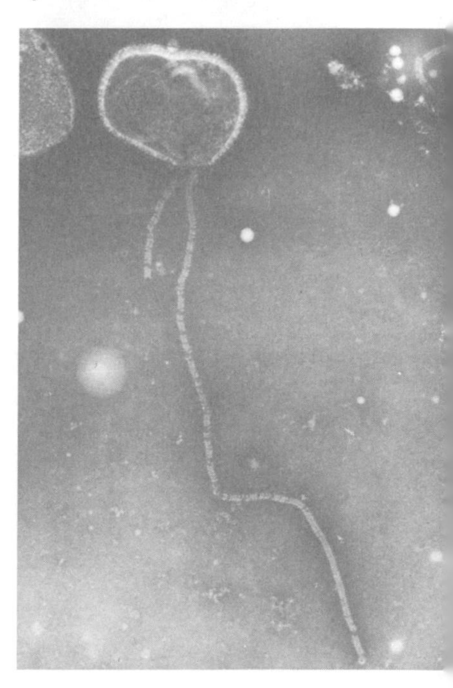

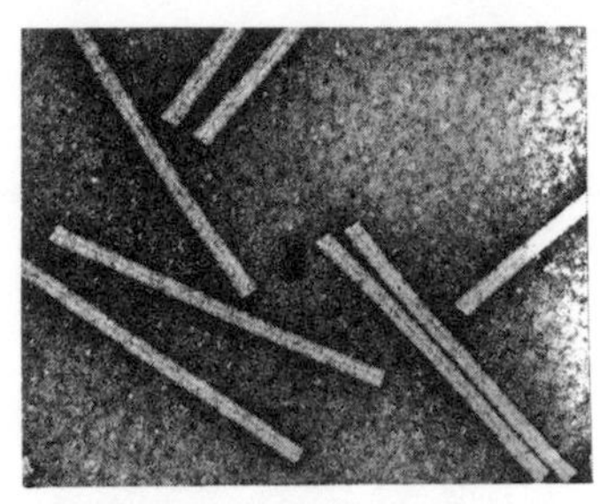

Above: a tobacco mosaic virus, which consists largely of protein molecules arranged around a central coil of ribonucleic acid
Below: an electron micrograph (enlargement: 450,000 times) and a laboratory model of the adenovirus, showing its regular 20-sided structure and some of its 12 projecting knobbed spikes

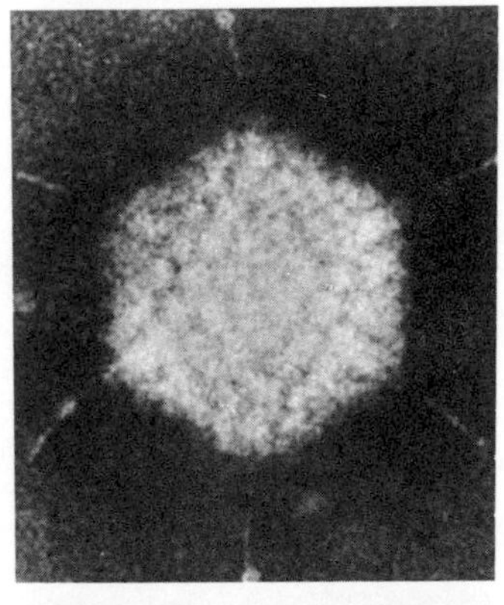

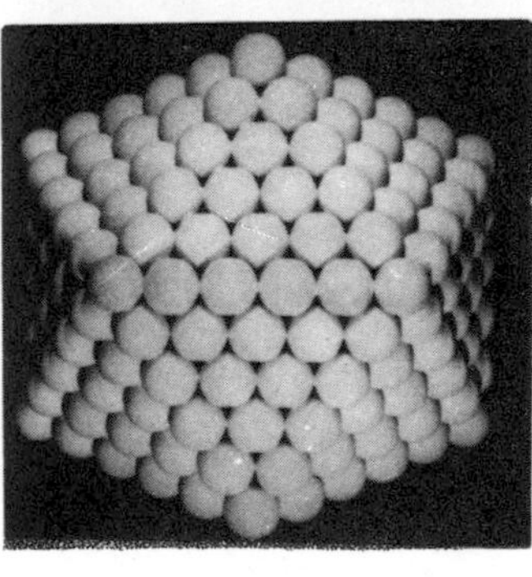

chromosomal DNA of mammalian cells. This type of incorporation of the virus DNA into the host DNA may well be one of the ways in which malignant cancer cells develop.

The virus which causes chickenpox often seems to settle in the cells of nervous tissue when the illness is over and remain in the body for many years without causing symptoms. Then quite suddenly it may reappear, causing the severe inflammation in the nerves and surrounding tissues which accompanies the very painful disease known as shingles. There is no doubt that the virus which causes shingles is the same as that which causes chickenpox, because if virus material is taken from a shingles infection and scratched into the skin of a child who has not had chickenpox, then that child will get chickenpox.

Small though viruses are, they can be studied under the electron microscope. To make the virus particles visible they are surrounded by an 'electron dense' staining material so that the virus is seen as a light, 'electron transparent' object on a dark ground.

Viruses with 'helical', 'cubical' and complex structures have been observed. The tobacco mosaic virus, for example, has a helical structure. It resembles a tube, the walls of the tube being made from protein molecules which sandwich between them a thread of nucleic acid. The virus is composed of 2,130 of these protein molecules – called capsomeres – with 16 of them to each turn of the helix. The viruses which cause influenza and parainfluenza have a similar structure, but they have in addition an outer envelope made from a mixture of material produced by both virus and host.

Many viruses which appear at first sight to be roughly spherical in shape, can, after detailed inspection under an electron microscope, be seen to be regular polyhedrons having what is called cubical symmetry. For example, the adenovirus, which causes a type of respiratory disease in man, is an icosahedron (a solid with twenty faces). It is made up of 252 capsomeres and has fibres sticking out from each of the corners. Many cubical viruses do not have as complicated a structure as this. The polyoma virus which causes tumors in rodents, for example, has only 42 capsomeres; while turnip yellow mosaic virus (which causes a disease in turnips) has only 32.

The T2 bacteriophage – another virus which infects the colon bacillus – has a remarkably complex structure: a large head containing the DNA and a tail with fibres on the end serving to attach it to the cell wall of the host bacterium. The tail contains an injecting mechanism – looking uncannily like a syringe – which helps to push the virus DNA into the bacterium.

Viruses are very varied, not only in structure, but also in the type of relationship they have with cells. There are some virus-like particles, called episomes, which may infect bacteria but do not multiply and destroy the host cell. In fact, episomes may perform various different functions useful to the bacterium, such as giving them resistance to antibiotics. There seems to be no hard and fast distinction between episomes which work for the cell and viruses which exploit or kill the cell.

It seems, then, that DNA viruses have evolved from the genetic material of other organisms – often from the DNA of the hosts they now prey upon. Similarly, RNA viruses must have evolved from the RNA messenger molecules produced by the genes of potential host cells. Indeed, viruses have been described as 'escaped genes'; and this is probably the simplest way of summing up most of what we know about them.

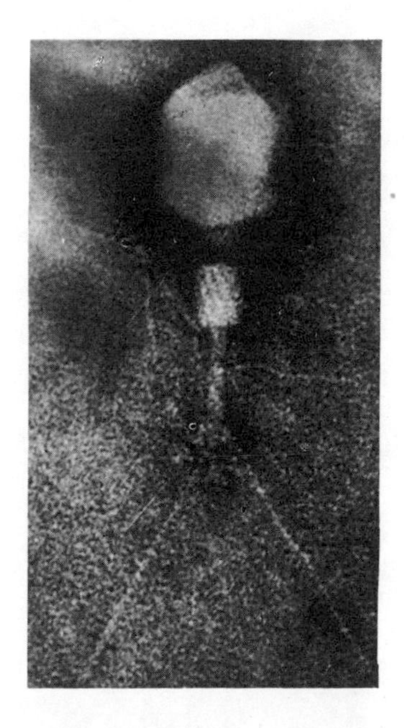

The T2 bacteriophage (above) adsorbed on to the cell walls of the colon bacillus (below left), and (below right) conspicuously present amid the debris of a lysed cell

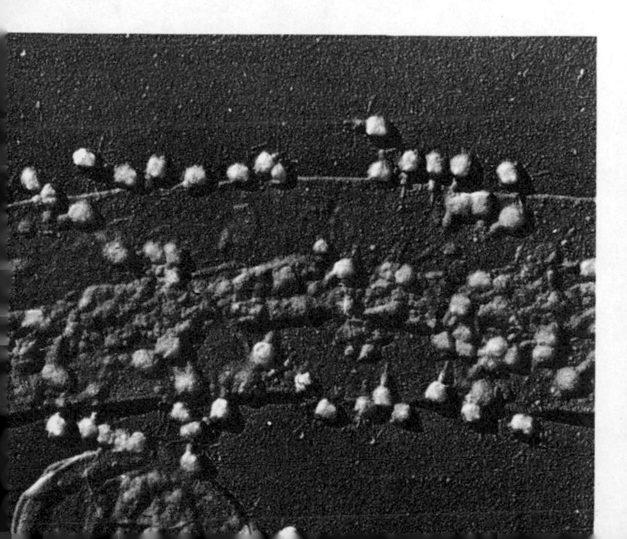

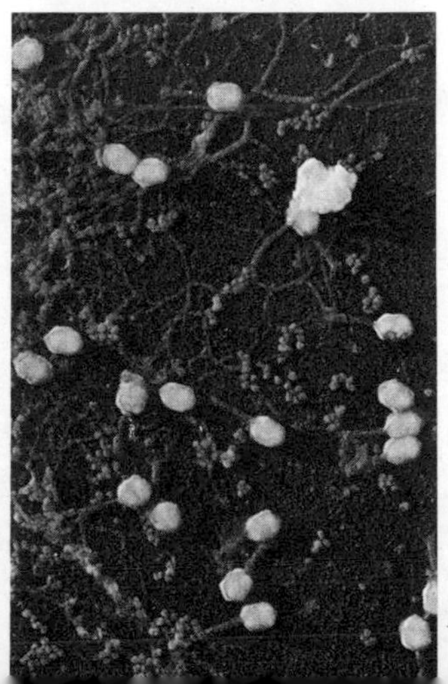

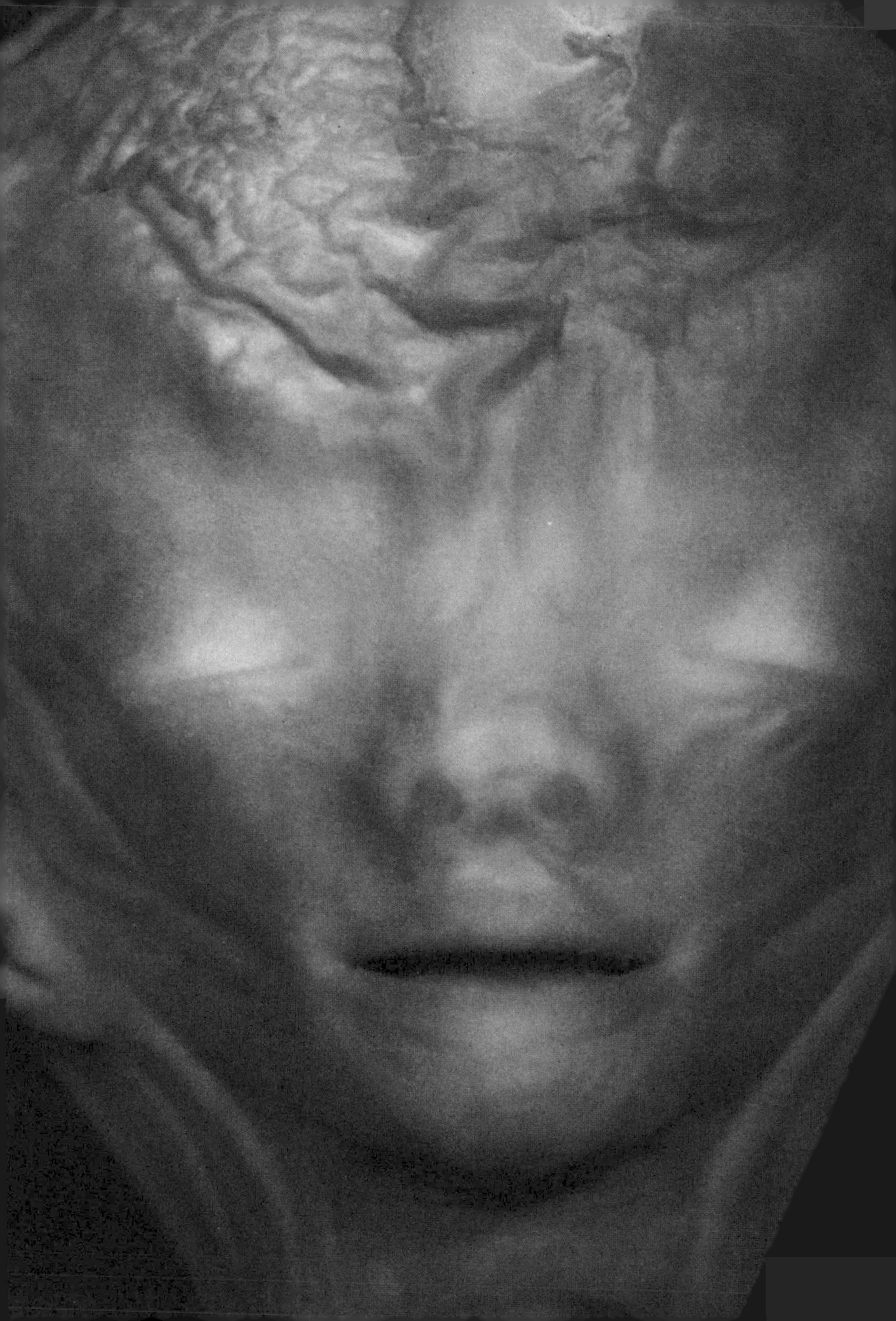

EMBRYOS AND CELL SPECIALIZATION 8

A fertilized egg cell is a very plain, structureless thing compared with the young creature into which it will develop. It contains all the organelles and chemicals that a normal cell contains and also a certain amount of yolk to feed the growing embryo. The nucleus of the egg, half of which is derived from the father and half from the mother, contains most of the information necessary for normal development to occur – although some essential information is also contained in the cytoplasm.

From egg to organism

To the biologist the true hen's egg consists only of the yellow yolk, and not the white, which is simply an additional food store for the growing chick. At the top of the yolk of a fertile hen's egg there is a little piece of cytoplasm containing the fertilized nucleus. When the nucleus begins to divide, cells are formed on top of the yolk which are not completely separated from the yolk beneath. To begin with these cells form a flat plate; this thickens and undergoes complicated changes which result in the formation of a chick embryo.

The frog's egg, which develops basically in the same way as a hen's egg, is about three millimetres in diameter and contains proportionally less yolk than a hen's egg. It is able to divide completely: first into two, then into four and eight cells, and so on. The yolk is concentrated in the bottom half of the egg and always remains at this end – named by the old biologists the

Opposite: the developing head of a human foetus – a vast assemblage of growing, dividing, migrating cells. At this stage, three to four months after conception, the whole embryo is still no more than six centimetres in length

'vegetable pole'. The top part of the egg – called the animal pole – contains less yolk and more pigment granules. The first division of the egg cell is stimulated by fertilization, but can also be triggered off by gently pricking the egg with a needle. In fact, egg cells which have had their nuclei removed can sometimes divide without being fertilized; so it seems that the process of cell division is completely under the control of the cytoplasm.

These early cell divisions lead to the formation of a bunch of cells called the blastula. At this stage the cells at the vegetable pole are much larger than those at the animal pole and contain much more yolk. They eventually develop into the gut of the embryo. A special portion of the egg where the yolky vegetable cytoplasm mixes with the pigmented animal cytoplasm goes on to form the embryonic head. However, if the animal and vegetable cytoplasm are artificially caused to mix in a different position by tilting the egg, then a different part of the egg can be made to develop into the head. This shows that the cytoplasm profoundly influences the development of the embryo.

If one of the two cells formed by the first division of the frog egg is killed, then the other cell develops into only half a frog. The eggs of insects behave similarly. The fertilized nucleus of the insect egg normally divides several times and the nuclei so formed migrate to various parts of the cell before the cell itself divides. But if any part of the cytoplasm of the insect egg is damaged – for example by pricking the egg with a needle either before or after the nuclei have finished migrating – then the adult insect will have some defect. In this type of egg the cytoplasm has an effect on development from the very beginning. There are, however, many animals whose eggs are able to compensate for early damage: thus if the fertilized human egg cell is accidently divided into two quite separate cells in the early stages of development, two perfect and separate individuals – identical twins – will be formed. The same is true of such animals as the sea urchin.

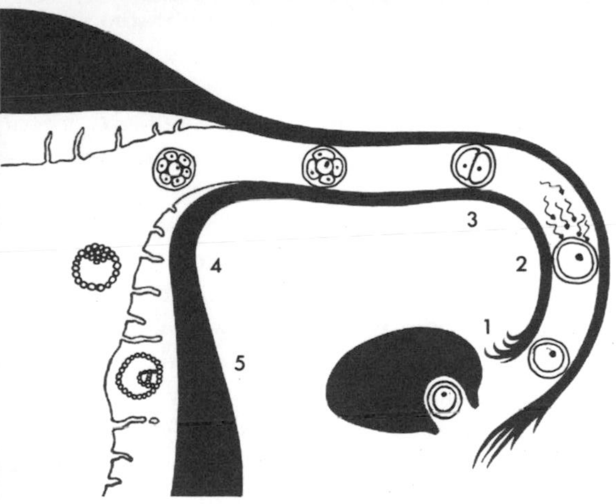

Above: the development of a frog egg from the four-cell stage to the formation of the gastrula. Left: the early development of a human egg cell, illustrated diagrammatically. At ovulation a follicle on the surface of the ovary bursts, releasing a mature egg which then enters the uterine tube (1). If it is fertilized by a sperm cell (as at 2), the egg begins to divide (3). Within a few days it enters the uterus, forms a blastula (4) and implants in the uterine lining (5)

In the next stage of development of the frog the cells of the blastula begin to move over one another and dent inwards to form the gastrula – a double-layered structure with a hole at one end. It is the cells around this hole – the blastopore – that go on to form the head of the embryo. These cells are known to have a remarkable influence on the development of the whole embryo, for if some of them are transplanted from one embryo to any part of another embryo, a tiny new embryo will develop at the point where the cells were transplanted. The cells of the blastopore lip have a special power to induce the comparatively unspecialized cells of the gastrula to form an embryonic head and body – even if a normal head is forming already. Cells from other parts of the gastrula, however, do not have the ability to influence early development in this way. This type of influence of one developing tissue on another occurs in many different ways in the older growing embryo.

Teeth rudiments develop in the mouth of the growing embryo when two separate tissues come into contact; and certain tissues transplanted to the mouth can be induced to form teeth, which otherwise they would never do. Elaborate experimental work has shown that in certain cases a single protein from a particular tissue is responsible for inducing another tissue to develop in a particular direction. There may also be some inducers which are messenger-RNA molecules. These are supposed to act by entering cells – almost like an infecting virus – and causing them to synthesize special proteins which change the direction of the cell's development. Protein inducers are thought to act rather like the repressor substances found in microbial cells; but there is little solid experimental evidence to show that they actually do. Explanations of development in terms of molecules are still largely hypothetical, but molecular mechanisms in addition to those found in micro-organisms will almost certainly be found in developing embryos. A favourite suggestion is that histone proteins, which do not exist in procaryotic bacteria such as the colon bacillus, may exercise some control over gene action in higher cells by masking some genes and unmasking others in response to signals given by hormones or inducer substances. Further insight into some of these problems has come from some very simple but telling experiments involving the giant single-celled alga *Acetabularia.*

Switched-on cytoplasm

When Joachim Hämmerling began experimenting with *Acetabularia* in 1926, he did not know that his experiments would provide vital clues to the action of the cell nucleus of higher organisms – clues which were at first overlooked by the molecular biologists, most of whom were working with microbes. He chose to work with *Acetabularia*, because the organism consists of a single cell with only one nucleus and reaches an overall length of 3-5 centimetres – near the record length for a single cell. The nucleus comes at one end of the cell, in the root, and at the other end there is a cap which varies in shape according to the species. Hämmerling cut off the root – containing the nucleus – of one species of *Acetabularia* and grafted on to it the stem and cap of another species. The caps of these two species had quite different shapes; but after some weeks each cap had acquired the original shape of the other. The formation of the new cap had involved the synthesis of new enzymes and of new cap material: clearly it was the nucleus which gave the orders and the cytoplasm which obeyed.

Some extraordinary results were to follow: Hämmerling found that, if the nucleus is removed from a growing cell before a cap has been synthesized, then it will grow; but on changing the conditions appropriately caps will form. This shows that the cytoplasm conbe kept under special conditions for weeks, and no caps will grow; but on changing the conditions appropriately caps will form. This shows that the cytoplasm contains very stable 'information' which can be used to synthesize enzymes for cap formation even when the 'information' is weeks old. These experiments are now interpreted as showing that the nucleus can produce messenger-RNA molecules which are highly stable. These stable messenger molecules are able to act in a controlled way so that the cell goes on synthesizing protein and performing all the other normal chemical reactions in addition to synthesizing a normal new cap. There is some DNA in the cytoplasm, but this DNA is restricted to the mitochondria and chloroplasts, and could not make the messenger-RNA molecules necessary for synthesizing a cap.

Similar experiments have shown that the cytoplasm of developing cells taken from the newt are also

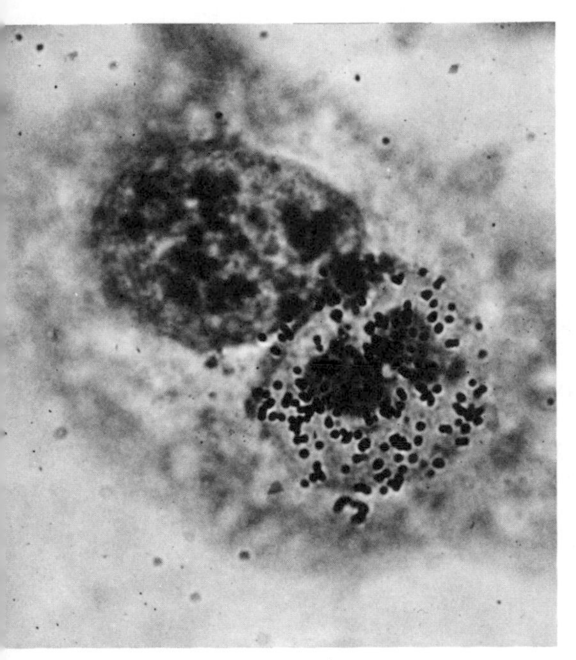

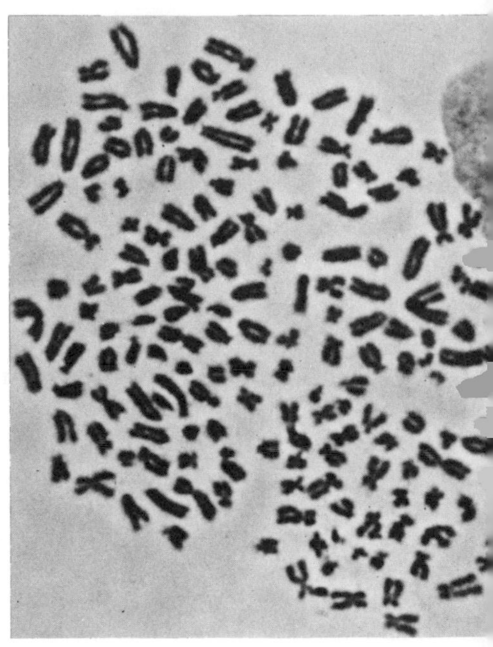

Man/mouse hybrid cells. Left to right: a cell containing a human nucleus (labelled with black grains) and a mouse nucleus which have not yet fused; the chromosomes of a single-nucleus man/mouse hybrid cell dispersed and arrested during cell division; and a small colony of man/mouse hybrid cells

switched on to their own special development programme. If cells destined to become pigment cells are taken from the newt before they have yet changed, and their nuclei are removed, they nevertheless develop, as destined, into pigment cells. This shows that their cytoplasm has already received instructions for future development – possibly special messenger-RNA molecules – before the nuclei are removed. Likewise the egg of the sea urchin can make repeated cell divisions after the nucleus has been removed; and amoebae, too, can survive a few days and synthesize proteins after their nuclei have been removed. The red blood cells of mammals lose their nuclei during the course of development; yet they still go on synthesizing haemoglobin afterwards. All these facts show that the cell need not be constantly dependent on the nucleus for everyday instructions.

In 1965 J. F. Watkins and Henry Harris produced the first hybrid animal cell by fusing the cells of two widely different creatures – man and mouse. They knew that

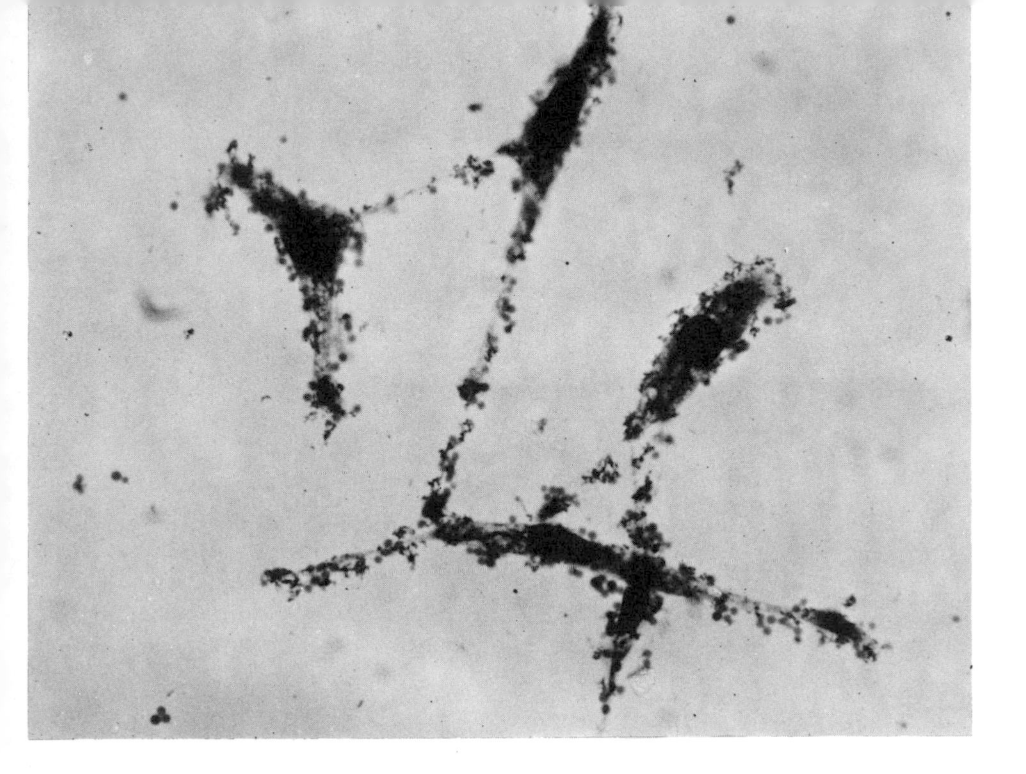

cells with more than one nucleus had often been observed in people who suffered from certain virus diseases. This suggested that cells might be made to fuse if they were mixed with a virus in a test tube. They performed the experiment – and it worked. It was not done with the ghoulish intention of rearing a man-mouse monster, but with the intention of making scientific studies of the relationship between nucleus and cytoplasm.

The virus used to fuse the cells simply has the effect of causing the surface of the cells to stick together so that several cells form a clump. Cells stuck together by viruses in this way soon develop cytoplasmic connections and eventually fuse. The virus particles are actually killed with intense ultra-violet light, lest they cause an infection in the fused cells. The nuclei of the hybrid cells at first remain separate, but after the first cell division they, too, usually fuse. Hybrid cells can live for a long time and can also multiply like normal cells. Although tissues cannot be transplanted from man to mouse without being attacked by the immune-rejection system of the body, and the eggs of mice cannot be fertilized by human sperms, cells from man and mouse can be made to fuse completely and grow as well as the cells from which they were derived. This shows that the activities of the cells of man and mouse must be very similar, although their bodies are so different.

The red blood cells of hens have small dense nuclei which do not synthesize DNA or RNA, or even divide, during their lifetimes. However, if these cells are fused with human cancer cells, the red cell nuclei both synthesize DNA and RNA and divide. The cytoplasm from the human cancer cell excites the dormant red cells into action. If, however, the red cells are fused with white blood cells from the rabbit, which are also non-dividing cells, healthy hybrid cells are formed, but they do not divide either. But the rabbit cells can be made to divide if, again, they are fused with human cancer cells. These experiments show that the quiescent nuclei of rabbit or hen cells will respond to something in human cancer cells which makes them divide. The experiment was first performed with cancer cells as a matter of convenience, but other types of dividing cells will also activate non-dividing nuclei after fusion.

The dense DNA in the dormant nuclei of non-dividing cells is reactivated by fusion with active cells, at the same time becoming more diffuse in appearance. Dense DNA is comparable to a tightly coiled ball of string; whereas normal DNA is present as loose dispersed loops, except when the chromosomes condense prior to cell division. Dense DNA is also sometimes found within active cells – often just one chromosome being affected. The mealybug, for example, has one set of dense chromosomes – the set inherited from the father – and another set of dispersed chromosomes inherited from the mother. This maternal set of chromosomes caters for all the needs of the mealybug – the paternal set being switched off. In female mammals there are two X chromosomes – the female is XX, the male is XY – and one of the two X chromosomes in the female is always dense and inactive.

It is possible in fact to show by genetic experiments that one of the X chromosomes of an animal is inactive. All cat lovers know that tortoise-shell cats – often called tabbies – are always female. Tortoise-shell kittens usually arise from the mating of a ginger tom cat and a wild, or dark-coloured, tabby. The tortoise-shell kittens which result have one ginger X chromosome and one wild X chromosome in every cell of their bodies. However, in each cell one or other of the X chromosomes has been made inactive. The tortoise-shell pattern of ginger and dark spots on the skin and fur show which chromosome is active, and where. The comparatively large patches of one colour are believed to be caused by the inactivation of chromosomes early in development when there are rather few cells in the embryo. Although it appears to be a matter of chance which chromosome is inactivated at this time, once inactivation has occurred all the daughter cells derived from the original cell in which inactivation occurred still have the same inactive chromosome. Why the X chromosome is inactivated in this way when both members of all other chromosome pairs are active is something of a puzzle. In women not all the genes on the inactivated X chromosome can be fully inactive, because women with only one X chromosome (Turner's syndrome) do not develop normal sex organs. However, mice with only one X chromosome are able to function as perfectly normal females.

The switching on and off of whole chromosomes and parts of chromosomes by changes in the coiling and density of DNA may partly determine the way in which cells develop and become specialized. However, the cytoplasm also plays a crucial role, as is shown by the experiments with hybrid cells and the experiments with *Acetabularia*. But the details of how *Acetabularia* synthesizes a new cap or of how the simplest changes of embryonic development occur are still a mystery.

Doubles by the dozen

Joshua Lederberg, awarded a Nobel prize in 1958 for his discovery of sex in bacteria, has said that we are on the brink of a major evolutionary perturbation. He was referring to new techniques which might eventually enable man to reproduce himself by a method described as 'taking cuttings' – a method which has been fully worked out in the frog.

The nucleus can be removed from the unfertilized egg of a frog and replaced with the nucleus of an ordinary body cell, whereupon the egg proceeds to develop quite normally, just as if it had been fertilized. Such experiments are comparatively simple to do in the early stages of development, when nuclei can be removed from cells quite simply: nuclei removed from frog cells at the blastula stage of development and transplanted into frog eggs from which the nuclei have been removed nearly always develop normally into frogs. Nuclei removed at later stages, however, do not so often develop into normal frogs but frequently die at the tadpole stage – though this may only be the result

(1) Apparatus used in the transplantation of the cell nucleus: a microdissector, a microscope, and a micropipette attached to a micrometer syringe. (2) A frog blastula cell – the donor cell – is sucked into the pipette and (3) injected into an enucleated egg cell. (4, 5) The two-cell stage and a more advanced stage of an egg with a successfully transplanted nucleus

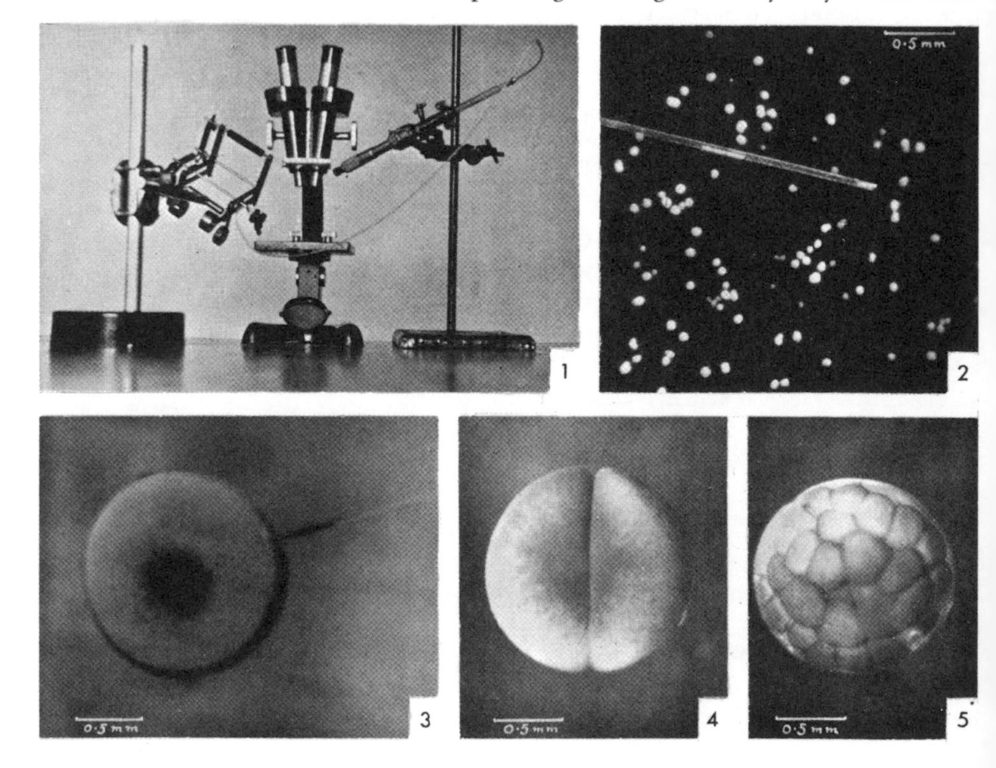

of the increased technical difficulties involved in using the smaller cells of the later stages. Normal frogs have actually been obtained by injecting nuclei from gut cells of tadpoles into egg cells from which the nuclei have been removed. And nuclei from a frog tumour have also been injected into frog cells, which developed into tadpoles before they finally died.

These experiments suggest that in frogs, and probably other organisms too, the nucleus is unaltered by the processes of development. Genes may be switched on and off during development, but the DNA itself must remain unchanged. There are, however, some organisms – insects and nematode worms – which show major changes in the chromosomes of the body cells during the course of development. In these organisms only the reproductive cells maintain the full number of chromosomes, the other cells losing chromosomes in the normal course of events.

Frogs developed asexually from cells belonging to a single individual by means of nuclear transplantation (cloning)

All the body cells of adult men and women normally contain the same number of chromosomes – and, so far as is known, no irreversible loss of genetic material occurs except perhaps as a result of chance mutation. In the future, therefore, it may be possible to take nuclei from the gut cells of one person and inject them into enucleated human eggs and so produce hundreds of genetically identical individuals. All the theoretical knowledge and much of the practical experience needed for this type of experiment already exists. If scientists really wish to do it, and the public permit it, people of this kind might be brought into the world within the next ten years or so. Eggs which have been injected with nuclei might be farmed out in the wombs of normal women or, more likely, in the wombs of women who would otherwise be sterile. Perhaps distinguished people with special abilities would be willing to donate a few healthy cells from their gut for this purpose: it would then be possible to produce genetic replicas of famous artists, musicians, politicians and scientists on an unprecedented scale.

It would also be possible to answer once and for all the problem known to Francis Galton as the 'nature-nurture' question. How many of the genetic replicas of an Einstein or a Churchill would grow up to be equally or even just moderately distinguished? At the present time, despite extensive investigations by geneticists, it is not possible to answer this question with any assurance. But those who imagine that they will be able to obtain brilliant children by means of egg transplants are, I believe, likely to be disappointed.

Migrating cells

Movements by which the cells change their positions in relation to one another are an essential part of the process by which the embryo develops. These movements bring together cells which may have originally been far apart. During the development of the frog, for example, after the gastrula is formed, a little ridge appears near the blastopore. This ridge gets bigger and extends backwards, until finally it curves over and fuses to form the nerve chord and the brain covered by skin. These external cell movements are followed by many more that occur inside the embryo during the formation of the various internal organs.

Cells seem to be able to recognize other cells similar to themselves, and this process of recognition becomes crucial when cells are moving and sorting themselves out. If the cells of a sponge are separated by pushing them through a fine gauze, they continue to live as separate cells; and if kept together, will once more stick to each other and form a sponge. The developing organs of frogs, which can be dissociated into individual cells by putting them into a solution free of calcium, will re-associate when calcium is returned to the medium. A piece of embryonic kidney can be dissociated and re-associated in this way: after dissociation all the cells become mixed up together; but when calcium is added the cells aggregate to form a ball and begin to sort themselves out. The cells which were originally on the outside of the kidney migrate back to the outside again and the cells which were previously on the inside migrate back to the inside. This shows that cells not only recognize each other, but are somehow able to sort themselves out by organized migration.

This process of cell recognition can be very sensitive. If the cells of two different species of sponge are mixed together, they will not form a single mixed sponge but two separate sponges, each containing only the cells of its own kind. Chick and mouse kidney cells mixed together however form a mixed kidney containing cells of both species, but, as before, with the inner and outer cells in their correct positions. One of the most remarkable examples of cells sorting themselves out is in *Hydra* – the fresh water anemone, belonging to the same family as the familiar sea anemone. If the body of *Hydra* is turned inside out, then the cells which now find themselves on the inside will migrate back to the outside where they belong, while the cells outside will migrate back inside.

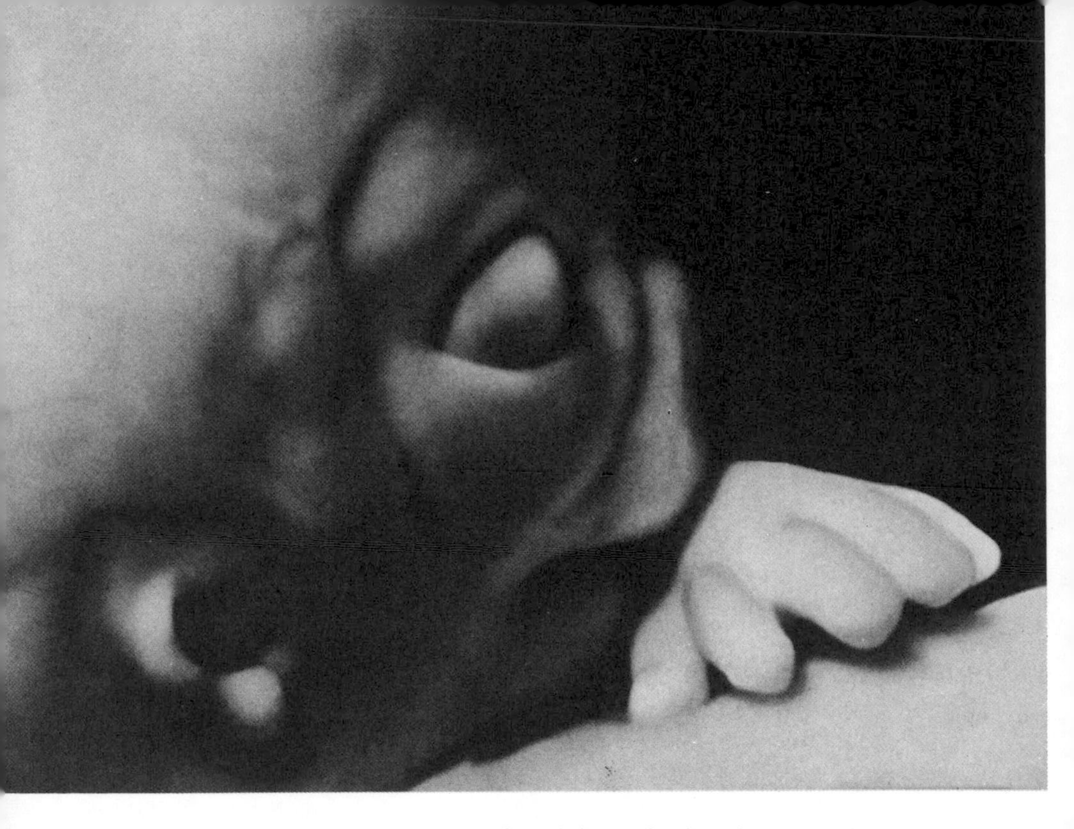

Stages in the development of eye and hand and of the cranial bones in the human embryo. The actual length of the photographically enlarged embryo illustrated above is no more than 2 cm; that of the embryo shown opposite is about 3 cm

This ability of cells of a particular tissue to stick together and recognize each other is believed to be due to some property of the cell surface. Loss of this property is one reason why cancer cells are able to migrate about the body and form secondary tumours in new places. Cells are stuck together by electrically charged atoms such as calcium and by a kind of glue which is a protein and carbohydrate mixture. Sometimes cells can be separated by simply removing calcium, but many cells can only be separated with the help of enzymes that digest the protein/carbohydrate glue.

Control by hormones

If more than half the liver of an animal is removed, the liver may yet return to its normal size. The human liver regenerates in this way if for some reason a portion is removed. And if one kidney is removed, the other kidney will grow until it has doubled its

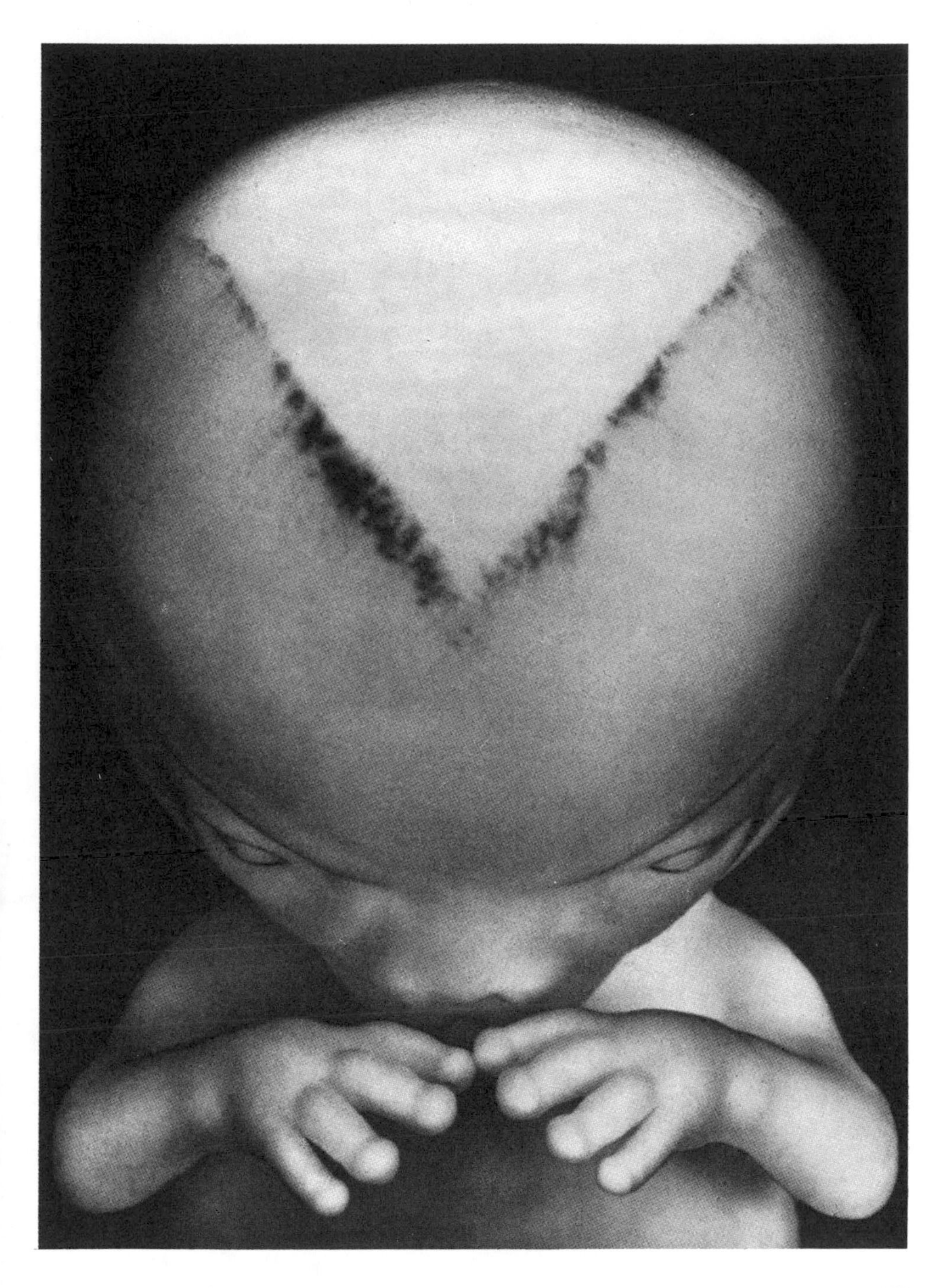

former size. It is not only in embryos that the growth of cells and tissues is constantly controlled – in adult organisms, too, each tissue and organ is maintained at a characteristic size proportional to the body as a whole.

The growth of organs seems to be controlled by some form of negative-feedback system involving hormone-like substances. Probably a hormone produced by the organ concerned increases in concentration until cell division in the organ stops. The concentration then begins to decrease as old cells die, and cell division starts again. There are also hormone systems which work by positive feedback. One of these controls the formation of red blood cells in the marrow. When the oxygen content of the blood is low, special cells in the kidney produce a hormone which induces the red blood cells to mature faster. When people go to high altitudes where there is less oxygen, these adjustments are made, resulting after some days in a higher number of red cells in the blood.

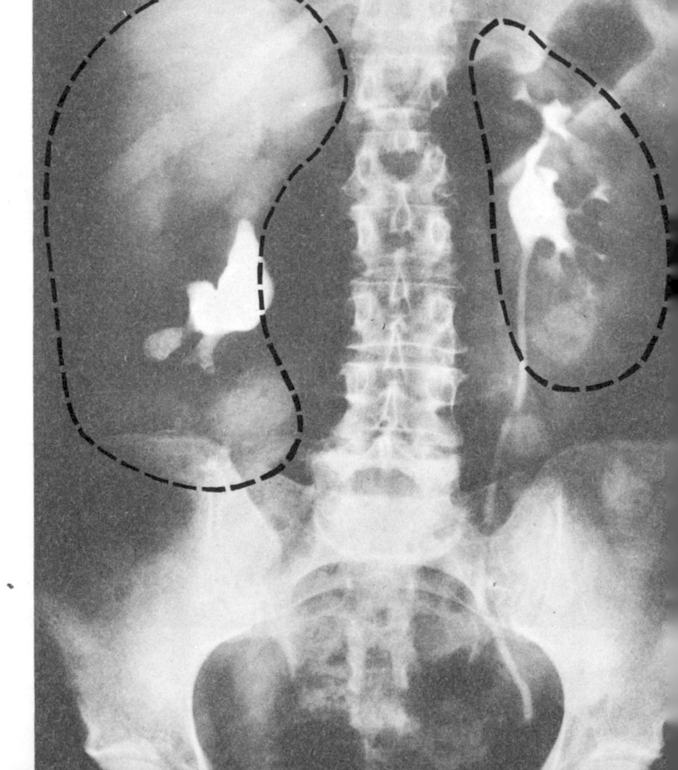

An X-ray photograph in which a pathologically enlarged kidney is compared with one of normal size. A healthy kidney, however, may compensate naturally for the removal of its diseased opposite by increasing in size to much the same extent as the larger kidney outlined above

No absolute distinction can be drawn between the more familiar hormones of the adult organism and the hormones, or inducer substances, which play a part in growth and development of the embryo. Hormones mostly seem to work by switching on genes; but they may also work by assisting enzyme reactions or by altering the properties of cell membranes. Oestrogen, a hormone produced by the ovaries which is responsible for the appearance of the secondary sexual characteristics of the female, is an example of a hormone which works primarily by switching genes on. If the ovaries of an animal such as a rat are removed and it is given oestrogen artificially the protein synthesized by the cells of the uterus may increase by as much as 300 per cent. If, however, the antibiotic actinomycin D is given at the same time, then no new protein is synthesized. Actinomycin D is known to prevent the synthesis of messenger RNA and so prevents previously inactive genes from becoming active and producing proteins. Oestrogen increases protein synthesis in most of the cells of the body and so must activate a great many different genes. In the hen oestrogen normally stimulates the production of yolk proteins in the liver, and if oestrogen is injected into the cock it will also make yolk proteins in its liver, which it would never normally do.

A little gland in the throat – the thyroid – has a profound effect on human and animal growth and development. Babies with abnormal thyroids often develop into cretins – mentally retarded dwarfs with large heads and misshapen hands and feet. The same gland plays a very important role in the development of the frog from the tadpole. In 1912 J. Gudermatsch discovered that young tadpoles could be made to develop into frogs before their normal time if they were fed with mammalian thyroid gland. Conversely, tadpoles from which the thyroid gland has been removed never develop into frogs.

The thyroid gland produces a hormone called thyroxine. And if this hormone is given to babies with abnormal thyroids it can prevent them developing into cretins. Similarly, if some thyroxine is given to tadpoles from which the thyroid has been removed the tadpoles will develop normally into frogs. The presence of this comparatively simple organic compound in the blood triggers off radical changes in the tadpole – the reabsorption of the tail, the growth of the lungs, the loss of the gills, the shortening of the intestine, and many other changes.

The amount of thyroxine in the body influences the metabolism of body cells. When the amount of thyroxine is high, the cells can do a lot of work: they burn up a lot of carbohydrate and fat and produce a lot of heat. When there is little thyroxine in the blood the body cells work more slowly and produce less heat – in fact cretins have an abnormally low body temperature. The amount of thyroxine which is produced by the thyroid gland is itself controlled by another hormone made by the pituitary gland. The thyroxine molecule contains iodine and it is essential to have sufficient iodine in the diet for thyroxine to be formed in proper quantities. In some parts of the world iodine is present in very small quantities in the soil and people living in these parts may develop swollen thyroid glands (goitre) and may suffer from thyroxine shortage – a condition characterized by swelling of the skin and general sluggishness.

Thyroxine acts directly on mitochondria, causing an increase in their activity; but it also directly affects the expression of genes in developing cells. The liver of the tadpole is induced by thyroxine to synthesize new proteins found in the liver of the adult frog and the tail is induced to digest itself by the activity of enzymes from the lysosomes. Tails removed from young tadpoles can even be induced to digest themselves in isolation, if they are placed in a solution of thyroxine.

Insects also have hormones which play a crucial role in regulating their development. The hormone ecdysone, for example, is responsible for initiating the change of the larva or grub of the fly into an adult. It can actually be seen to have a direct effect on the chromosomes of certain insects having chromosomes large enough to be observed directly under the microscope. These chromosomes are puffed up in places where a lot of RNA is being synthesized – the puffs are in fact genes in action producing messenger molecules. When ecdysone is injected into larval insects, puffs appear at particular places on the chromosomes: when small quantities are injected only one puff appears; but when large quantities are injected a second puff appears. During normal development, as the concentration of ecdysone in the larva slowly increases, first one then another gene is activated before the larva changes into the adult.

Hormones have a wide variety of different types of chemical structure, ranging from comparatively simple organic molecules to proteins. Many seem to act by influencing gene activity; but how they do this is not known. It seems quite likely that they may act in a similar way to the small molecules – the inducers – which induce enzymes to form in micro-organisms.

Although hormones act directly on cells the end result involves the whole organism. However, hormones not only have a physical effect, they can also have a psychological effect. Mood and certain aspects of personality are dependent on the quantity of various hormones in the blood. Women commonly notice a change of mood occurring at the different stages of the menstrual cycle and develop special feelings when they are pregnant or when they breast feed their babies. These moods are induced by changes in the hormones circulating in the blood, although purely psychological causes are also at work. The thyroid gland also has profound effects upon temperament. Overactivity of the thyroid gland produces anxiety and rapid fluctuations of mood; underactivity produces mental sluggishness or apathy.

Congenital abnormalities

The genes provide a blueprint for development which, if all goes well, will produce a genetically perfect multicellular organism. Unfortunately, development

A young mongol child (facing the camera) at a school in Belgium for mentally handicapped children

is sometimes abnormal, and imperfect young are brought into the world only too frequently. About one in a hundred children born have some sort of serious abnormality and a larger number have minor defects. Mercifully many more abnormal babies are lost as miscarriages and abortions.

Abnormal development may be the result of inherited defects or be caused by virus infections or drugs affecting the mother during the early stages of pregnancy. These three causes of abnormalities act in a broadly similar way: they interfere with the normal growth of cells in the embryo or with the normal expression of genes in these cells. One of the most commonly found defects is mongolism, which is now known to be caused by inheriting, as a result of an abnormal cell division, one small extra chromosome in every body cell. Somehow this small extra chromosome interferes with the normal action of the other chromosomes in the cell; and the result is that mongols develop abnormalities of the face, eyelids, tongue and other parts of the body, including the lines and creases on the palms of the hands and soles of the feet. Mongols are also mentally subnormal, although they usually have a very pleasant nature. Many other abnormalities of development are known to result from extra chromosomes or even from whole extra sets of chromosomes, such that an embryo may begin to develop with $3n$ or $4n$ chromosomes per cell, instead of the normal $2n$ chromosome complement. Most of these embryos are aborted – or, in many animals, reabsorbed – long before full term.

Women who get German measles in the first twelve weeks or so of pregnancy run a 10–25 per cent chance of having a baby with abnormalities of the eyes and internal ears causing deafness and partial blindness. The virus, which may cause a very mild or scarcely noticeable illness in the mother, passes through the womb into the tissues of the baby, where it may cause a severe illness which interferes with the normal growth of the delicate tissues that are being laid down in the first few weeks. Other common virus diseases, such as influenza, are also believed to cause abnormal development of babies in a small proportion of mothers in the first three critical months of pregnancy.

The side-effects of some drugs taken during pregnancy have been tragically illustrated in the case of thalidomide. This drug, if taken early in pregnancy, prevents the normal development of the limbs. Many beautiful and intelligent children were born with no arms or legs, or with pathetic stumpy limbs, because their mothers had been prescribed this drug. Thalidomide seems to interfere directly with the expression of certain genes, thereby preventing the development of parts of the growing limb buds.

Thalidomide was not properly tested on animals before the drug was marketed, otherwise the dangers would have been detected. Many other drugs which are commonly used were brought into use before the dangers of drugs to pregnant women were fully realized; thus it is most unwise for women to take any drugs, other than those prescribed by the doctor, in the first three or four months of pregnancy. Recent

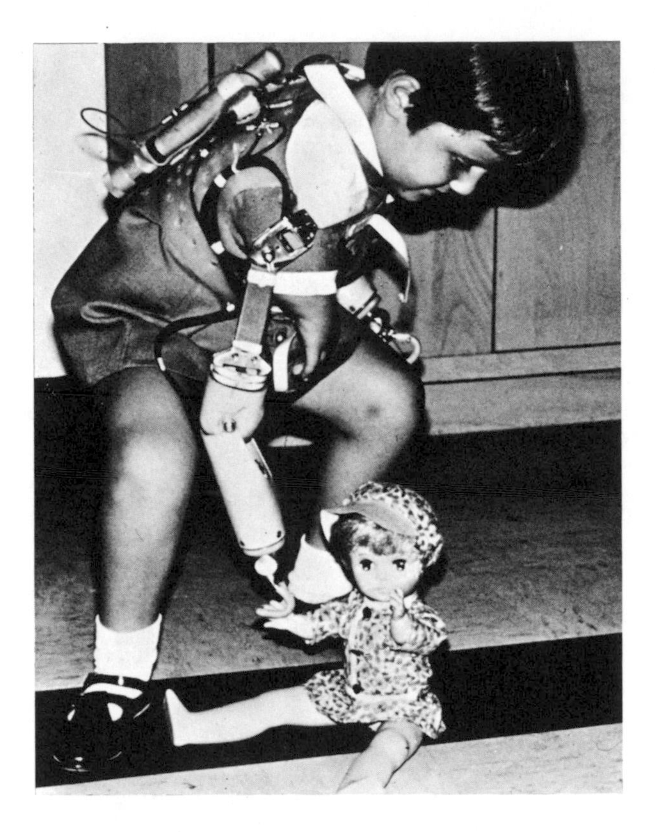

Inadvertently rendered armless while still an embryo by the effects of the drug thalidomide, a young girl fitted with artificial arms stoops to gather up a doll

research has shown that smoking during pregnancy causes the baby to develop more slowly and so to have a lower weight at birth. Smoking is not believed to cause any abnormalities, but is almost certainly undesirable. Several claims have been made by scientists that the drug LSD causes mutations and also abnormalities of development in experimental animals. Other scientists have, however, obtained results which do not support this conclusion. It is possible that LSD is harmful in some animals and not in others, but it is also possible that LSD does not have any harmful effects in people and that the preliminary results were misleading. Further research will have to be done. Although science usually gets the right answers in the long run, in the short run, it can be wrong, particularly if there are openings for prejudice.

Muscle and nerve cells

The cells of the body could be described in many different ways. There are the cells, of the liver, the kidney and the intestines, which perform the biochemical work of the body; the cells, of the heart and the other muscles, which do the mechanical work; and the cells, of the nerves and brain, which control the body in a voluntary and involuntary way. Some cells, such as the red cells of the blood and the keratin cells of the surface of the skin, consist almost entirely of one protein – haemoglobin in the former case and keratin in the latter. Other cells, such as those of the liver, make a large number of different proteins, store carbohydrates and act as a general chemical factory. The way in which cells perform their biochemical work has already been covered in general terms and it is not possible to go into much more detail in this book. However, I will say a few things about the operations of two highly specialized types of cell – nerve and muscle.

Muscle cells contain nuclei and mitochondria and all the other structures found in normal cells. The so-called voluntary muscles – those we are able to control by will – consist of many cells joined together to form fibres. In each fibre there are many nuclei, but they have no cell membranes in between. Under the microscope these fibres can be seen to consist of light and dark stripes – and it is these that have provided clues to the way in which muscle cells work. A dark stripe with one half of a light stripe on either side of it forms a contracting unit which is 0·0002 centimetres long – the basic muscle. Under an electron microscope these individual bands can be seen to be formed of even smaller filaments: the dark bands of thick filaments and the light bands of thin filaments. The thin filaments of the light band run in among the thick filaments, but stop before they reach the centre of the dark band. When the muscle contracts, the light bands disappear as the light filaments slide across and into the dark filaments. The filaments slide across each other in the same sort of way as two halves of a pack of cards being

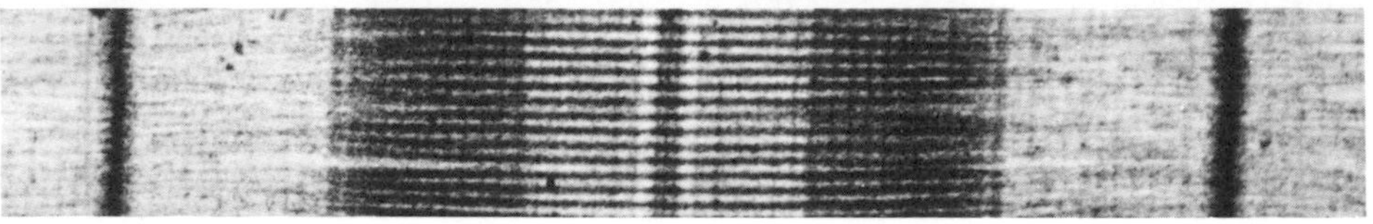

Striated muscle tissue from a rabbit is enlarged 16,000 times (top). Each of the diagonal 'ribbons' is a thin longitudinal section through a muscle fibril and consists of a series of contractile units, one of which is further enlarged (above). Below: a diagram of the contractile mechanism, showing how the overlapping thick and thin filaments give rise to the characteristic appearance of striated muscle

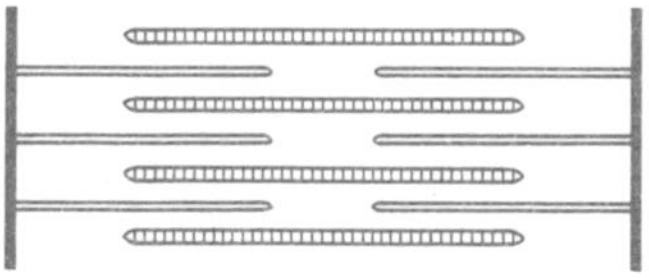

pushed together, after the person who has shuffled them has flicked them into position at the corners. It is quite different from the contraction of a coiled spring.

Muscles are very largely protein and the filaments which slide across each other as the muscle contracts are composed almost entirely of two different pure protein molecules. They are long fibrous molecules which are able to combine together. If these two proteins are squeezed together through a fine tube they form a long spaghetti-like thread which will contract if it is placed in a solution of ATP (the energy-providing molecule) and suitable salts. Neither of these proteins will behave like this if they are separated from each other. Exactly how they are able to slide across each other by using the energy from ATP is not known; but there are cross-bridges between molecules of the proteins which might conceivably act like a ratchet.

The nerves are the cells that transmit messages from the brain and spinal cord to all parts of the body. They are rather like electric cables, with a live portion on the inside covered by an insulating sheath. However, nerves are not like the cables of the domestic electricity supply that carry power to lights, heaters and household gadgets: they merely carry a signal which switches on the muscles – the muscles themselves doing the work. The system works rather like an automatic device for controlling a television set. The viewer sits comfortably in a chair, a cable and control switches enabling him to change from one channel to another and to control volume and brightness. The viewer controls the picture, but the power for the set comes directly from the mains. In a similar way the brain controls the body.

Parts of the nerve cell are very elongated and it is these parts that form the nerve fibre and its sheath. The part of the nerve cell that contains the cell nucleus is always located in the brain or spinal cord, from which the fibres stretch out to various parts of the body.

Below left: a section through a small region of the cerebral cortex. Right: a single neuron, picked out by a special process of staining with silver

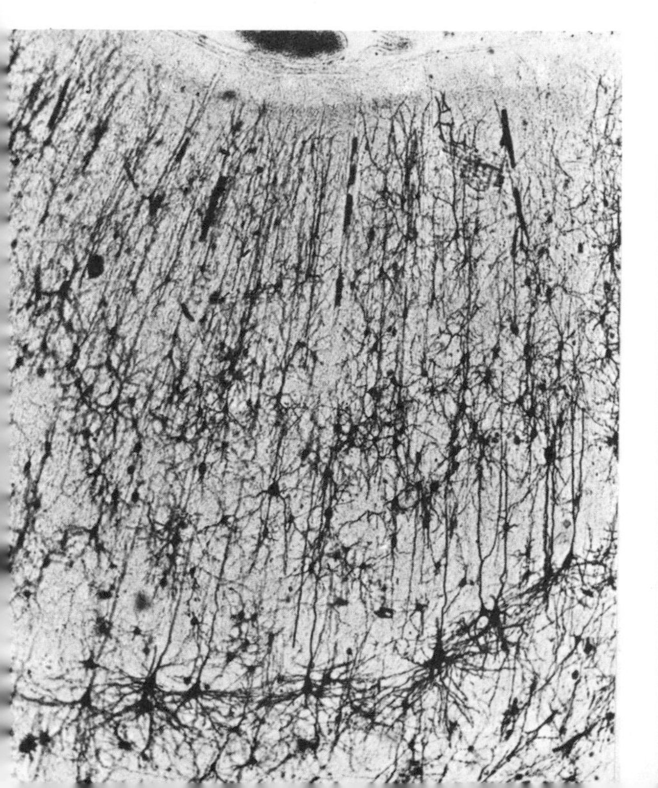

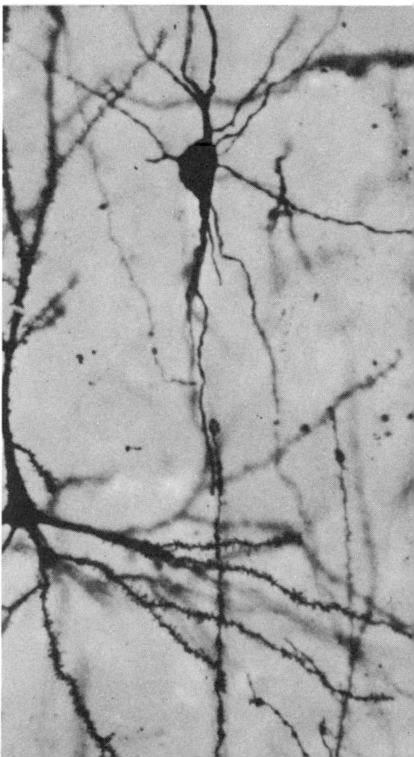

The Italian physiologist Luigi Galvani discovering, in 1771, that a frog's leg suspended by the nerve from a brass hook kicks convulsively if the toe makes contact with a silver plate. Without knowing it Galvani was demonstrating not only the generation of an electric current but also the electrical nature of the nervous impulse

Nerve fibres make connections – the synapses – with other nerve fibres and with muscles. The many synapses in the spinal cord enable messages to be relayed to and from different parts of the body. The pathways of some of the simple messages through the nerves and spinal cord have been worked out, but the details of many more-complicated nervous pathways may never be known.

Stephen Hales, rector of Teddington in Middlesex in 1730, did some of the first clear-cut experiments showing nerve pathways in the frog. He suspended a frog – in which the brain had been destroyed – by its fore legs, and then showed that if the toes of the hind legs were gently squeezed the frog would withdraw its foot. The two nerve messages that played a part in this response go in and out of the spinal cord at different points. A message is relayed from the foot to the spinal cord, entering the latter by nerves towards the back. This message then passes across one or more synapses, activating another nerve cell, which transmits a message towards the front of the spinal cord and thence out to the muscles of the foot.

This circuitry can be quite simply established by cutting the nerves in various places and studying the effect of this on the system. Simple nervous responses of this kind – the reflexes – are found in nearly all animals, although it is possible for some of these reflexes to be inhibited by the action of the central nervous system: if someone knows that their toes are going to be tickled or pinched they may decide to try to ignore it; but if the tickling comes as a surprise they will almost certainly withdraw their foot by reflex action.

Quite a lot is now known about how messages are passed down nerve fibres, but we are still very far from understanding how the brain really works. The electric message is carried along nerves as a result of changes in the quantities of potassium and sodium salts inside and outside the cell. Outside the cell there is a lot of sodium salt and little potassium, whereas inside the cell there is a lot of potassium and little sodium. This has an effect something like that found in a wet battery where different solutions separated by a semi-permeable membrane give rise to an electric current. If a very fine tube is inserted inside an individual nerve fibre and another tube is bathed in the external liquid, an electric difference of 65-95 millivolts can be measured between the two points.

When a nerve is stimulated, sodium flows into it at the point of stimulation and potassium flows out, discharging the local voltage difference across the membrane. The voltage difference is then discharged in the next bit of membrane, a little farther along, and so on – thus the message travels along the nerve. The message cannot go backwards, because the bit of membrane immediately behind the message is always discharged and it takes a few milliseconds for the membrane to return to its original state. The return to the original state is brought about by a so-called 'sodium pump'. Experiments at Cambridge using a giant nerve fibre (of a squid) nearly a millimetre in diameter have shown that the sodium pump uses ATP as a source of energy. There is also a system of enzymes involved in the process but, although this is a subject of very active research, the full details are still far from being known.

The brain contains thousands of millions of cells which relay messages from one part to another. The connections between all the nerves are far too detailed to be worked out in full, although most of the major connections are understood. In recent years there has been a great deal of work centred on the problem of discovering a biochemical basis for the process of learning. Some psychologists, impressed with the progress of molecular biology, have postulated that the brain's basic memory store might subsist in the RNA molecules in its cells. The idea was that information might be stored in the base sequence of these molecules in the same way in which genetic information is stored in DNA and RNA. But this suggestion raises as many scientific problems as it solves and it cannot possibly account for short term memory which is too rapid for a biochemical process of this kind. The advocates of the theory conducted experiments using rats and flatworms which had been taught to perform various tricks. RNA was extracted from the brains of the animals which had learned these tricks and injected into 'naive' animals which had not learned any tricks. The naive animals were then tested to discover whether they would now be able to perform the tricks which they had never learnt. Many experiments of this kind were performed and positive results were claimed in many cases. However, later more careful experiments have produced negative results, and the idea that memory has some basis in RNA molecules is now regarded by most scientists as being unlikely to be true. A great many scientists are currently engaged in studying the biochemistry of the brain, but there are no signs of a breakthrough as yet.

Cells and seeds

Opposite: embryonic material in flowering plants, as sketched by Matthias Schleiden in 1838

The germination and growth of seeds, a constant source of joy and satisfaction to the gardener, is often considered a boring commonplace by others

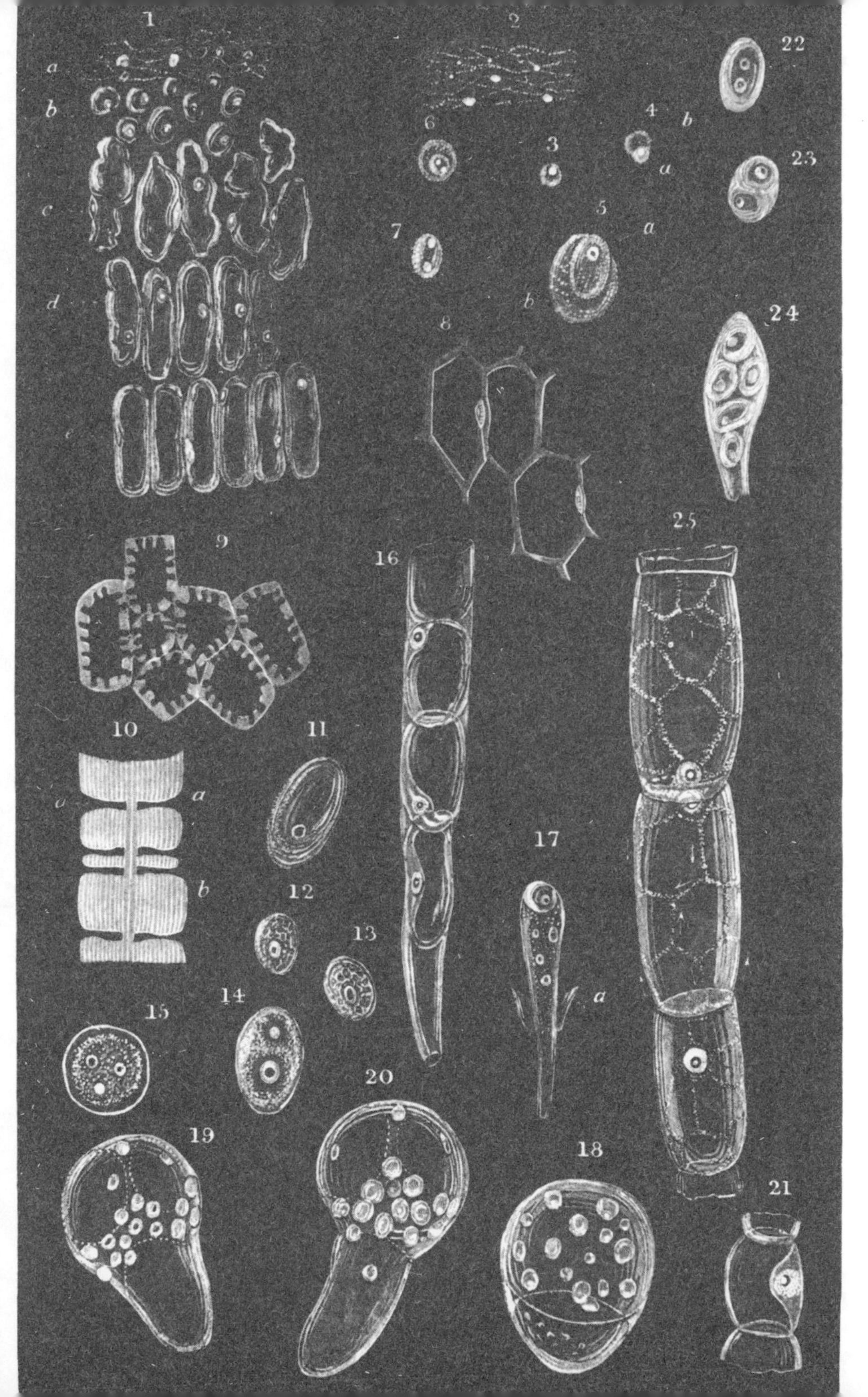
1
2
22
a
b
b
4
6
3
a
23
c
5
7
a
d
b
8
24
e
9
25
16
10
11
a
a
17
12
b
13
a
14
15
20
19
18
21

who remember germinating peas in sawdust and blotting paper at school. Higher plants, however, although they do not have as many different kinds of cells and tissues as higher animals, present similar problems to biologists studying how different tissues arise.

The seed from which a plant grows is a complicated multicellular structure. It arises from the division of a fertilized egg cell in the female part of the plant. Pollen cells, formed by reduction division on the male part of the plant, are carried by wind or insects to the female organ. A pollen cell landing on the female organ grows forming a long tube which extends towards the egg cell. The nucleus in the pollen cell divides and two haploid nuclei move down the pollen tube, until one of these pollen nuclei fuses with the haploid egg nucleus to give a fertilized egg cell.

The young plant is formed from the egg nucleus, which divides to form the growing germ layers of the root and shoot. When the seed germinates the specialized cells of the root and shoot are formed, later the specialized cells of the leaves, and finally the flowers. Layers of unspecialized growing cells are left behind in various parts of the plant. New buds or rootlets form from these growing cells in ways determined by hormones. Unspecialized growing cells are also found in the trunks of trees and are responsible for the thickening of the tree trunk, which occurs throughout its life. Once the specialized cells have been formed from these growing cells they do not normally divide again. However, some remarkable experiments with plant cells have shown that under special circumstances specialized plant cells can often revert to the embryonic type. Root cells taken from a carrot can be made to grow into a normal carrot plant with stem and leaves by growing the cells in coconut milk. If small pieces of carrot root (the part we normally eat) a few millimetres in diameter are gently shaken in coconut milk the small pieces of carrot begin to increase in size and divide. Individual cells are knocked off the main piece as a result of the shaking and grow

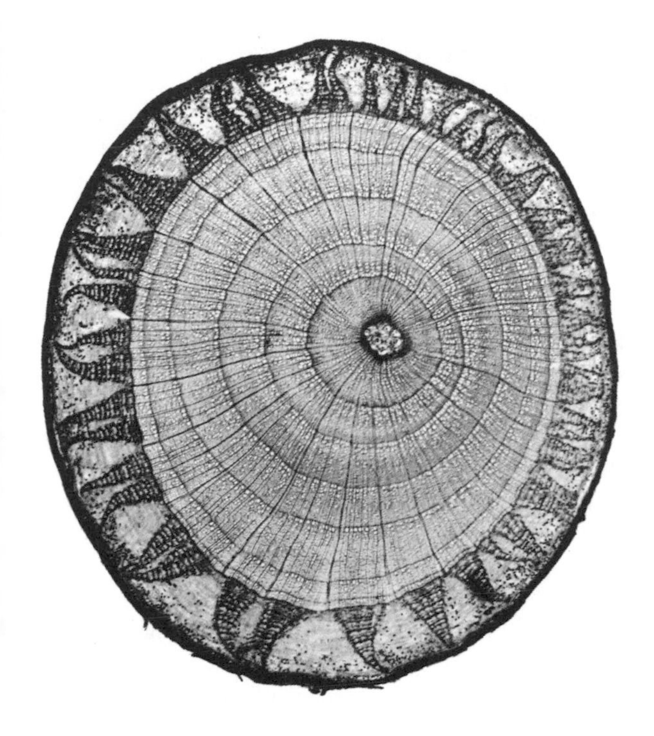

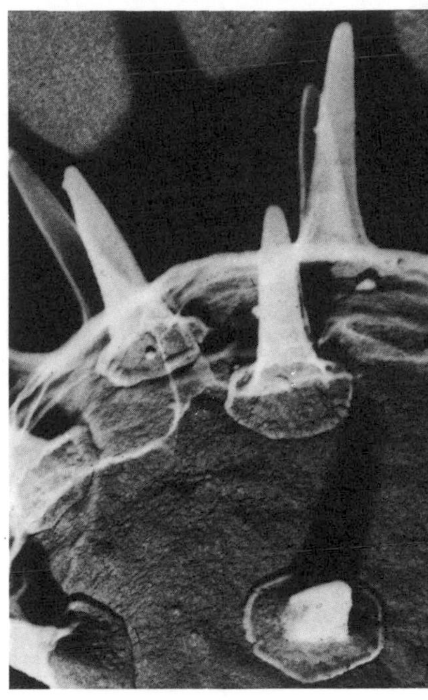

Above left: section through the stem of a linden tree, showing the structure of the bark and the annular rings in the woody core. Right: the spore of a fungus

independently. Some of these cells develop into little plant embryos with miniature roots and, if transplanted on to a solid medium free from harmful micro-organisms, will grow into a healthy green carrot plant. Under these circumstances the specialized root cells have realized the full potentialities of an egg cell.

It might be said that the secret of this experiment lies in the coconut milk. The coconut milk is the liquid food store of the embryo coconut tree and it contains several hormones which are able to switch on the genes of the carrot root cells so that they grow like an embryo. These plant hormones are found in other seeds besides the coconut, which is merely a very convenient source from which they can be obtained. Techniques of this kind for growing plants from single cells have already been put to practical use as a method of obtaining virus-free plant strains. Virus-free strains of potato produced by means of this method at Rothamsted experimental station in England have resulted in greatly increased yields of potatoes with enormous financial savings.

A young student, Fritz Went, working in his father's laboratory in Utrecht, Holland, in 1923, was the first person to discover plant hormones. He found that if the growing tip of a plant is cut off and placed on a small block of gelatine a substance diffuses into the gelatin block which can be shown to affect the growth of seedlings in other experiments. If the gelatin block is placed on the top end of a decapitated growing seedling, the seedling can be made to grow to one side or the other, depending on which side of the growing end the gelatin block is placed. This experiment was the first to show that substances are produced by the growing tips of shoots which affect the growth of cells distant from the tip: these substances are the plant growth hormones, the auxins.

If the leading shoot of a plant, for example the topmost shoot of a fir tree, is removed, then one of the other shoots below will take over and become the leading shoot. The whole shape of a plant is governed by the balance of the hormone concentrations produced by the leading shoot and the side branches. Large concentrations of the hormone will inhibit the growth of shoots and will stimulate the growth of stems, whereas with lower concentrations the side shoots may grow more. In fact there are several plant hormones now known and the control of plant growth is much more complicated than this. The auxins which have these effects have been purified and identified chemically – one of the most well known being a simple organic chemical called indole acetic acid.

The discovery of plant hormones has been of the greatest economic importance and has initiated a new era in the chemical control of crops. A simple example is in the preparation of plant cuttings. If the cut end of a twig is dipped into a solution of growth hormone it will quickly grow roots; in this way cuttings can be taken from trees, such as cocoa, which are otherwise very difficult to propagate vegetatively. Other plant hormones are used to induce flowering and to prevent unripe fruit from falling from the tree. Some fruits, such as the tomato and the fig, can be induced to develop without being pollinated merely by spraying with plant hormone; this is not only extremely convenient for some farmers but produces a seedless fruit. Weed killers such as 2,4-D are chemically related to indole acetic acid, but are effective growth inhibitors at very low concentrations. Indole acetic acid stimulates growth at low concentrations but inhibits growth at high concentrations. The 2,4-D inhibits the growth of broad-leaved plants much more than the growth of grasses and cereals and so can be used to keep down agricultural weeds in cereal crops as well as to kill lawn weeds.

A disease of rice, called foolish seedling, first found in Japan, is caused by a fungus which makes the plants grow much taller than normal. In 1926 E. Kurosawa isolated the active substances from the fungus. These substances were called gibberellins after the scientific name of the fungus *Gibberella fujikuroi*. The gibberellins are a larger and more complicated organic molecule than the auxins, but they have a similar overall effect on growth of shoots, although they inhibit the growth of roots. Gibberellins are also economically important; in particular they are used to increase the number and size of grapes and prevent rotting. Auxins act on the cell by stimulating cell elongation but not growth. Gibberellins act by actually increasing the number of cell divisions in a tissue. It is not yet known exactly how these plant hormones act at the level of the cell. There is some evidence that they act by switching genes on and off but this has not been established yet.

Man is slowly establishing dominion over the plant world, and there is no doubt that by the use of chemicals and plant hormones greater control will be established in the future. This, together with the possibilities of breeding plants to satisfy any requirement of taste, shape or appearance, is already bringing to the garden and the table living things designed if not created by man.

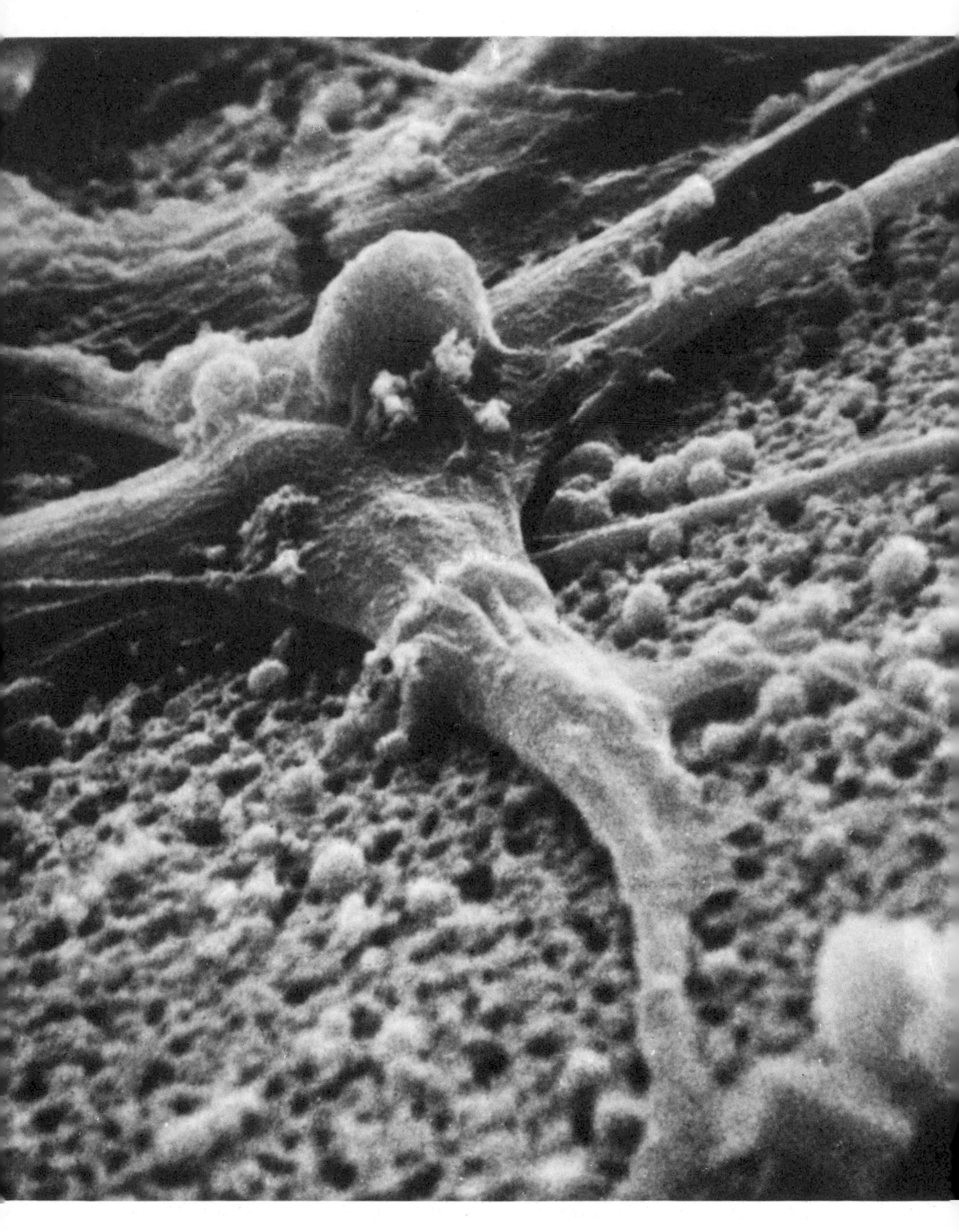

WHEN CELLS GO WRONG 9

Every type of living cell is normally adapted to a particular set of environmental conditions, although some cells are able to tolerate change better than others. The cells of multicellular animals and plants will only grow independently of their parent organism under very special conditions in the laboratory. On the other hand, the cells of micro-organisms may adapt to a wide range of habitats and, within a particular habitat, deal with fluctuating conditions. To a greater or lesser extent, all cells have the means to exploit certain kinds of environmental changes and the defences to buffer themselves against the worst effects of others. Even so, ionising radiation and chemicals can damage the cell DNA; while viruses may attack the cell and kill it, or change a normal cell into a malignant cancer cell. Cells and organisms protect themselves mainly by producing antibiotic substances designed to kill other cells and in higher organisms by producing antibodies designed to kill micro-organisms.

Genetic mutation

Although it is a remarkably stable molecule, DNA is vulnerable to damage by ultraviolet light, ionising radiation and various chemicals, including cellular enzymes. All cells consequently mutate at a low rate all the time. Most of these mutations are harmful, but a few may be advantageous, and organisms bearing these mutations may eventually outnumber other

Opposite: a scanning-electron micrograph of a cancer cell. Comparison of such studies has revealed that cancer cells move – and so invade healthy tissue – by means of wavelike motions in their cell membranes

similar organisms, especially if the population containing them is confronted with a change of environment. This low rate of mutation – the spontaneous mutation rate – is caused by cosmic radiation and radiation from radioactive molecules in food and in the environment, and partly by the hazards inherent in the replication of the DNA itself. Sometimes the wrong bases may accidentally be paired during DNA replication, leading to a permanent change in one of the daughter cells. Some chemicals which cause mutations do so by interfering with the replication of the DNA molecule; but others act indirectly by, for example, interfering with DNA synthesis or with the supply of bases.

Ultraviolet radiation from the sun, X-rays, atomic radiation, radioactive fallout and natural radiation from rocks are all harmful to cells. Ultraviolet radiation is very easily absorbed by most materials and in animals does not penetrate the superficial layers of the skin. White-skinned people are, however, liable to get malignant skin cancers if their skin is exposed to brilliant sunshine over a period of many years. These skin cancers, called melanomas, are caused by the ultraviolet light mutating the cells of the skin which make the brown melanin pigment. The mutated melanic skin cells cease to be subject to the normal controls of the body and multiply to form a black tumour.

A cancer of the skin, in the region of the eye

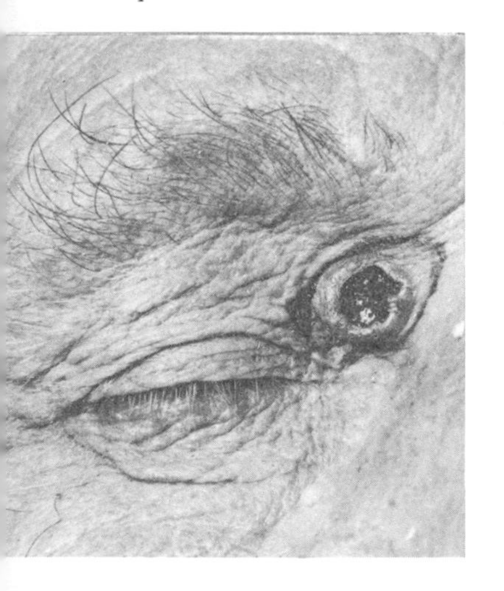

Ionising radiation, such as X-rays, penetrates biological materials and damages the nucleus and the cytoplasm of the cells. Amoebae with irradiated cytoplasm but unirradiated nuclei live for up to three weeks, whereas amoebae with irradiated nuclei and unirradiated cytoplasm live for only three days. The process of DNA synthesis is very sensitive to irradiation, so dividing cells and continually growing tissues, such as bone marrow, are the most sensitive to radiation.

Genetic damage may be caused by comparatively low quantities of ionising radiation. Ionising radiation is measured in Röntgen units (r); a dose of about 3r might be expected to be received by one person over a period of about thirty years, although this will depend very much on where he lives and whether he has had any medical X-ray examinations. Medical X-rays of the head, teeth or chest give a dose of only a few thousandths of a Röntgen to the sex organs, whereas X-rays of the pelvic region give doses of 2-3r to the sex organs. Experiments with mice have shown that a dose of 1r to the sex organs produces about 25 easily recognized mutations per 100 million genes. In *Drosophila* the rate of mutation is 15 times less for the same dose and the rates for other organisms vary even more widely. However, if we can assume that human cells behave like mouse cells, it is possible to say that a 10r dose of ionizing radiation given at high intensity to both of the future parents of each of a hundred children will cause five of the children to carry a mutation. These mutations will usually be recessive and so may go unrecognized for generations – until two recessive mutations of the same gene coincide in the same person.

Recessive genes in man behave in exactly the same way as the recessive genes in the peas studied by Mendel. Parents who both carry the same recessive gene may expect one in four of their offspring to show the effects of the gene. Most mutations, however, being harmful, cause the young organisms in which they occur to die at an early stage of development. Mutations are generally harmful because the genetic constitution of any organism has evolved over millions of years and so any sudden change in the DNA is liable to upset a delicate balance.

The rate of mutation induced by radiation can be measured experimentally in mice, but it is much more difficult to measure in man. Studies have been made of radiologists who were exposed to higher-than-normal amounts of radiation in the course of their job, and also of people who were exposed to radiation when the atomic bombs were dropped on Hiroshima and Nagasaki. However, the number of malformations that occurred in the children of parents exposed to radiation could not be shown to be greater than the number

The detonation of the first hydrogen bomb, at Bikini in November 1952

occurring in the children of unexposed parents. The only detectable difference was that larger numbers of female children were born to exposed than to unexposed mothers. Although the statistical significance of this ratio change is questionable, it could be taken to indicate that more mutations occurred in exposed mothers. A change in sex ratio would be expected with an increase in the mutation rate because recessive genes on the X chromosome, which are most likely to be lethal, are expressed in males, who only have one X chromosome, but not in females, who have two.

It has been estimated that fallout from nuclear bomb tests so far conducted will lead to an increased exposure of up to 0·5r or more over a thirty year period. This is less than the amount of radiation received by many people for medical reasons, but the two situations are not strictly comparable, because radioactive isotopes may be accumulated preferentially in certain organs of the body, especially in the bodies of growing children.

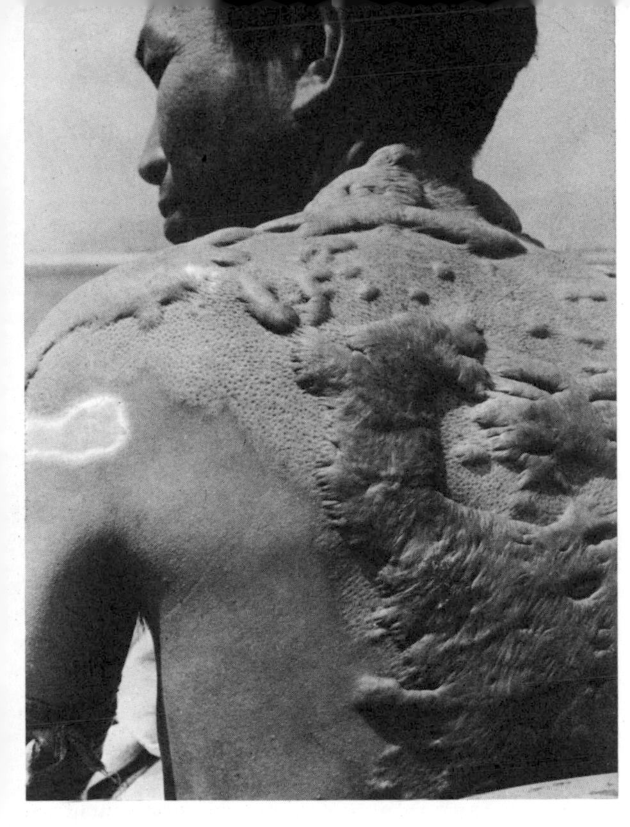

Radiation burns suffered by a survivor of the atomic explosion at Hiroshima in 1945

Ionising radiation does not only cause genetic damage. A dose of 100r to the whole body will lead to radiation sickness and a dose of 400r or more usually leads to death. A dose of 400r would be received at a distance of about one mile from the centre of an old-fashioned atomic explosion. Madam Curie, who first purified radium, and many other scientists and medical men, suffered from radiation burns and radiation sickness in the early years of this century, before the dangers of radiation were appreciated. The symptoms of radiation sickness are vomiting, fever, bloody diarrhoea, loss of hair and haemorrhages in the skin. A survivor of radiation sickness usually suffers from permanent scarring of the skin, and also, after a delay of some years, from various forms of cancer – particularly leukaemia (cancer of the white blood cells).

Some protection against radiation is possible – substances such as cysteine and cysteamine have a protective effect if administered before the radiation dose is received – but complete protection is not possible. These radio-protective substances act by mopping up reactive molecules – the so-called free radicals – which are produced as an indirect effect of radiation. The cell itself is also able to make good some of the damage to the genetic material by means of repair enzymes which cut out damaged sections of DNA and replace them, using the undamaged sister strand as the template.

A damaged cell is also in danger of being mutated by its own lysosomal enzymes, which may leak out of the lysosome sac and begin to digest the chromosomes. If the cell survives such an episode it will almost certainly have acquired some mutations. Chemicals may mutate the DNA of the cell either directly by altering the DNA, or indirectly by, for example, first unleashing the lysosomes. The first chemical found to cause mutations was mustard gas. Charlotte Auerbach, a German refugee working in the Institute of Animal Genetics in Edinburgh, made this discovery in 1939, but it remained classified until the end of the war, when it was found that German scientists had made the same discovery. Since then many more chemical mutagens have been discovered, and many of these have been used in experimental work to confirm the analysis of the genetic code.

A mutation may involve a very small change of just one base in the DNA message or perhaps a more complicated set of changes affecting several bases. However, the commonest types of mutation involve either a change in a single base or a complete break in the DNA strand. If two of the DNA strands break, the broken ends may rejoin the wrong way, creating two new chromosomes consisting of the parts of the old ones. This does not always kill the cell and is an important factor in the evolution of new species.

Typical effects of radiation on the chromosomes of a dividing cell – in this case a lily cell

Cancer – cells gone wild

Cigarette smoking is known to cause cancer of the lung; the inhaling of asbestos fibres can also cause lung cancer; mineral oils may cause cancer of the scrotum; hydrocarbons, such as 3:4-benzpyrene, can cause

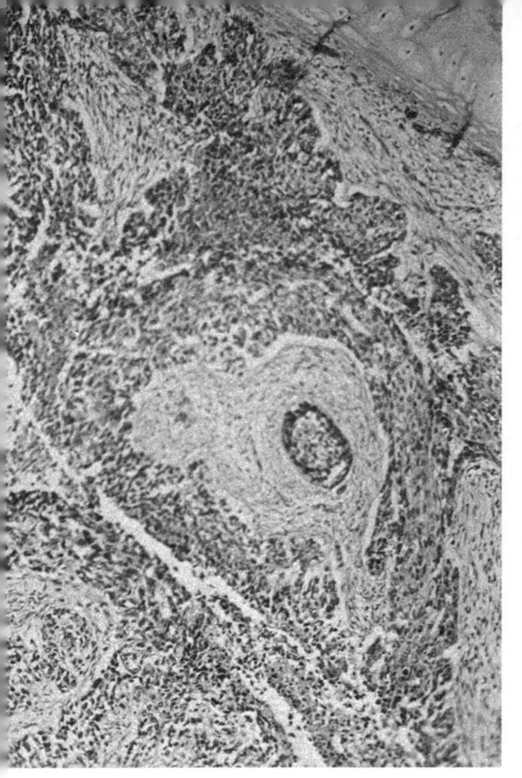

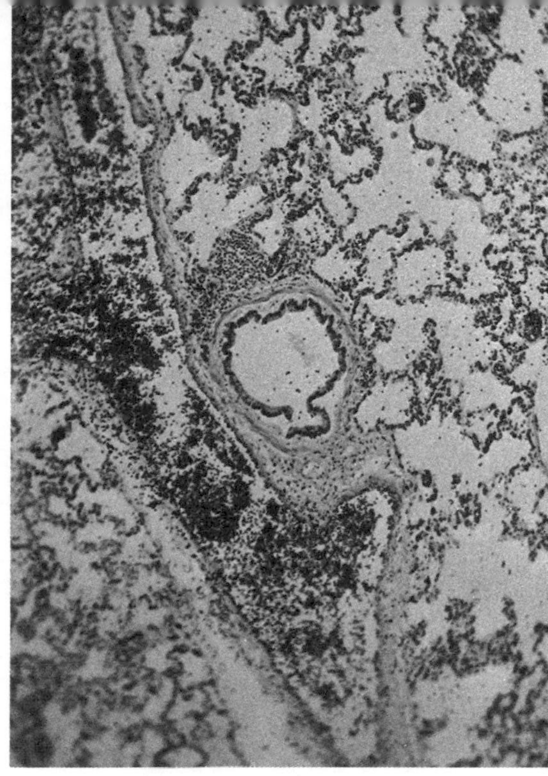

Normal lung tissue (left) compared with cancerous lung tissue

cancer of the skin. This list could be extended very much farther, but it would be misleading if it gave the impression that chemicals or noxious substances are the only or even the primary cause of cancer. Viruses may cause cancer in man, as they have been found to do in animals, but this has yet to be demonstrated conclusively – scientists are limited in the kind of experiments they can do with human beings. Human warts are tumours caused by viruses, but warts are never malignant – they are not cancerous. However, most scientists working in the cancer field expect to see it proved within a few years that viruses cause certain human cancers. Though it seems equally likely that genetic mutations are also a cause of some cancers.

When a cell has become malignant it grows relatively quickly and individual cells may spread via the blood stream and lymph ducts to other parts of the body to form more secondary tumours. Although the growth of cancer cells is abnormal in that they do not remain localized in tissues as other cells do, there is no

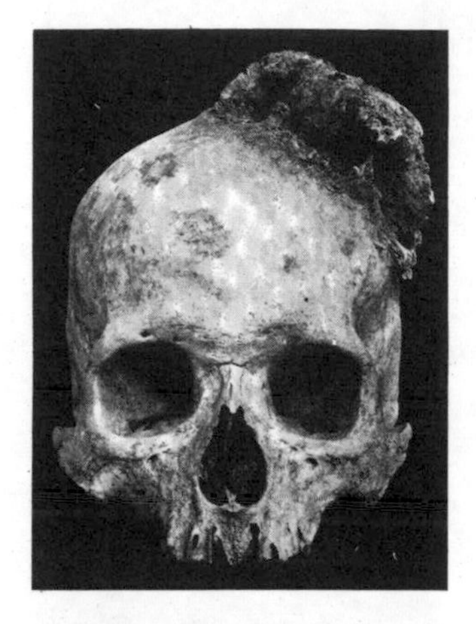

A cancer of the cranial bones

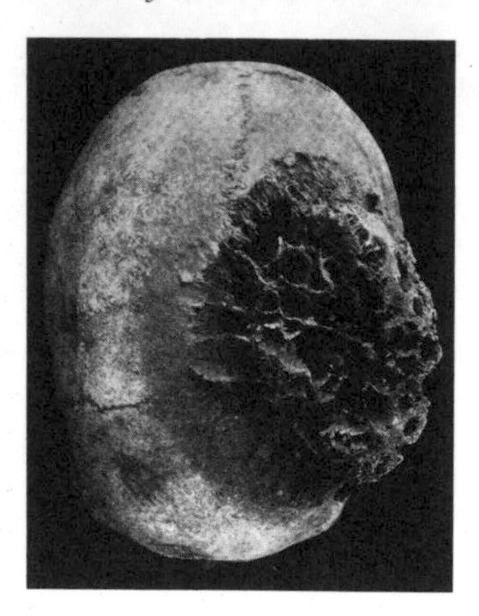

simple biochemical difference between cancer cells and normal cells. Cancer cells often rely more on anaerobic metabolism to supply energy than do normal cells, but this sort of difference cannot be said to account for the characteristic behaviour of cancer cells.

In 1910 Peyton Rous made his first observation leading to the discovery that a particular cancer of chickens – which came to be called the Rous sarcoma – was caused by a virus. Fifty-six years later, at the age of 87, he was awarded a Nobel prize. During this time evidence slowly accumulated, as a result of the work of Rous and many others, that viruses may be a common cause of cancer in animals; and many viruses are now known which, though they do not normally cause cancer in nature, can cause cancer under special laboratory conditions. Viruses known to cause cancer in animals as a result of artificial infection include a number of RNA as well as DNA viruses. Some of the DNA viruses have been found to have a structure and base composition resembling the DNA of their hosts, which suggests that they may really be escaped genes.

In the case of at least one cancer-causing virus it has been shown that the virus DNA becomes incorporated into the DNA of the host cell – in much the same way as the DNA of some bacterial viruses can become incorporated into the DNA of the bacterium. The virus can then multiply in step with the host cell; but the host cell is no longer the same – it begins to produce new proteins under the influence of the virus. The body usually makes antibodies to the new proteins, which are recognized as being foreign. The antibodies assisted by white blood cells then attack the tumour. Sometimes this may cause the tumour to regress spontaneously, and in healthy people abnormal cells are probably destroyed continually as they arise.

Even when the cancer shows signs of establishing itself as a recognizable tumour the body is still fighting it. One of the problems of cancer therapy is to give a treatment which will kill the cancer cells without killing the white cells, which play so important a part in the body's defences. Most of the drugs used to kill cancer cells also kill the white cells and the antibody-producing cells of the bone marrow. The best results are obtained when the cell-killing drugs can be applied locally to the tumour. Once the cancer has spread and secondary tumours have arisen the chances of killing the tumours by drugs or of removing them by surgery are very poor.

Hormones have been used very successfully to treat certain cancers. These are the cancers of the breast, womb and prostate – tissues which are strongly influenced by hormones in their normal state. But not all cancers of these tissues can be treated successfully by hormones. In some cases treatment is successful to begin with, but then the cancer cells seem to mutate and no longer respond to hormone therapy.

As long ago as 1896 doctors had observed that removal of the ovaries often caused regression of breast cancer, although it was not then known that the ovaries made female hormones. Hormone therapy of breast cancer is complicated and the principles upon which it works are not well understood. The treatments used for women before and after menopause are different and may involve removal of the ovaries, removal of the adrenal glands (sources of the hormones adrenalin and cortisone), removal of the pituitary hypophysis and the administration of male hormones, female hormones or cortico-steroid hormones. Why some of these treatments work is a mystery: sometimes female hormones are used in one stage of treatment and male hormones in another and yet both treatments may be effective in making the tumour regress at the time of use. Treatment for cancer of the prostate is much simpler and very effective – the testes are removed and female hormones are administered.

Hormone therapy often involves side effects – masculinization in the case of women and feminization in the case of men – these are the direct results of the hormones given and can only be regarded as the lesser of two evils. Fundamental knowledge of cancer has increased greatly in recent years, but methods of treatment are still very limited. Advances in knowledge of

A brain tumour, in which specially administered radioactive mercury has become localized, shows up on this radiograph as a cluster of dots

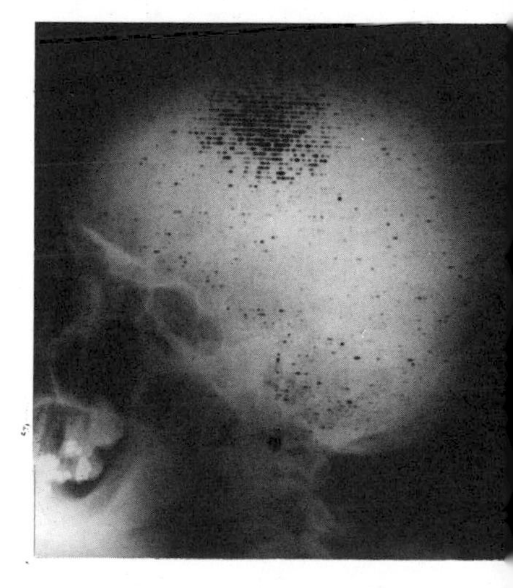

the processes of virus infection and the immune defences of the body are likely to make an even greater contribution to our understanding of cancer in the future.

Inherited chemical defects

In 1902 the English physician A. E. Garrod published a paper in *The Lancet* drawing attention to a rare disease called alkaptonuria. People who have this disease excrete large quantities of a substance called homogentisic acid in their urine, turning it black. These people are quite normal and healthy, mentally and physically, although they are more likely to develop osteoarthritis than normal people. Garrod pointed out that people suffering from alkaptonuria either had the disease or did not have it – there were no intermediate forms. He also pointed out that the disease was found in a much higher proportion among the relatives of those with the disease than among the general population, and concluded that he was dealing with an inherited defect in metabolism rather than a disease caused by a micro-organism. This view was quite novel at the time, as the great successes of bacteriology had led many people to believe that bacteria were the basis of all disease.

People suffering from alkaptonuria are unable to break down homogentisic acid to simpler substances as normal people do, so must excrete it unchanged. Garrod saw that this disease was one of a general type which he called 'inborn errors of metabolism'. In modern terminology these diseases would be described as being caused by a gene mutation leading to the loss or partial loss of an enzyme activity. Such mutations are usually recessive and only show up on the average in a quarter of the offspring of two carriers of the gene.

Another similar disease, called phenylketonuria, is the result of the secretion of phenylalanine in the urine. People suffering from this condition are usually feeble minded, the high concentration of phenylalanine in the blood affecting their brains. They lack an enzyme necessary to convert phenylalanine into tyrosine – part of the same biochemical pathway as is blocked in alkaptonuria. It is interesting that defects of two closely related enzymes should have such vastly different effects on the whole organism, one leading merely to mildly incapacitating osteo-arthritis and the other leading to feeblemindedness.

Many inherited metabolic diseases also have effects on the function of the brain. Recently Dr Ida Macalpine, a physician, and Dr Richard Hunter, a psychiatrist, have suggested that many members of European royalty suffered from a disease called porphyria and that this disease has been responsible for the psychological peculiarities of some of them. The families of the European royal houses have intermarried a great deal and there are, of course, full records of their pedigrees together with medical records of individual monarchs going back to the sixteenth century. Porphyria may have been the cause of the 'madness' of George III and George IV, and of the life long illnesses suffered by Mary Queen of Scots, James I (of England), Frederick the Great and others.

English monarchs James I (left) and George III, probable sufferers from the hereditary disease porphyria

People suffering from porphyria make increased quantities of porphyrins in the liver and secrete them in their urine, which becomes coloured dark red. The porphyrins are made as a result of the breakdown by enzymes in the liver of the haem portion of haemoglobin molecules. Other symptoms of porphyria are severe abdominal pain, vomiting and constipation, and often the skin is very sensitive to sunlight. Severe attacks of the disease may occur every few months or at intervals of many years. The disease is caused by a dominant gene which does not always express itself in every generation. In modern times about 12 per cent of those who suffer from porphyria have been certified as 'insane'.

Another inherited disease is haemophilia (bleeding disease). Queen Victoria carried a gene for haemophilia and passed it on to at least two of her daughters and one of her sons. The gene for haemophilia was not known in the royal family before Victoria and probably arose as a mutation in the previous generation. Haemophiliacs lack enzymes in the blood that cause the blood to clot. Before the advent of modern medicine they used to die of uncontrollable bleeding from a small wound at an early age. Nowadays small amounts of blood from normal people, containing the clotting enzymes, can be given to haemophiliacs by blood transfusion. The condition is caused by a sex-linked gene which is carried by the females but not expressed in them. The gene is expressed in the sons who get it and, if they live, they pass the gene on only to their daughters, who will all be carriers.

Errors of metabolism may also be inherited. Modern medicine can offer help in dealing with many of these inherited metabolic disorders. For example, some babies cannot metabolize the sugar galactose. If they are given a galactose free diet from an early age they will develop normally; but if they are brought up on a diet including normal amounts of cow's milk, their brains may be irreversibly damaged. Similarly, infants with phenylketonuria can be given foods containing a minimum of phenylalanine and so feeblemindedness can be avoided. Diabetes, another disease, which is sometimes the direct consequence of a genetic defect, can be controlled by injections of insulin. Perhaps other genetic diseases will some day be treated by the regular injection of enzymes to break down such substances as homogentisic acid or phenylalanine, which otherwise would damage the body or the brain. Other possibilities, which have been given the futuristic name 'genetic engineering', may include the injection of normal DNA carrying the instructions for making enzymes needed by a person with some defect. A benign virus might also be used to carry desirable human DNA into all the cells of the body.

The introduction of pieces of useful human DNA in conjunction with virus DNA is the most plausible idea, but we are still a long way from putting it into practice. The provision of normal DNA to abnormal cells could only be of help in certain kinds of genetic defect where irreparable damage has not yet been done. Damage which has occurred during the course of development may be too great for enzymes or DNA to rectify. It is, for example, difficult to see how mongolism could ever be cured by either of these techniques.

Perhaps one of the most likely developments in the near future is the injection or transplantation of normal cells to make up for the defective function of certain body tissues. Haemophilia, diabetes and sickle-cell anaemia might all be cured by transplants of this kind once the immunity problems have been solved on a practical scale, and there now seem to be no insurmountable theoretical blocks to prevent this from being done.

The effects of drugs

The first man successfully to bring a scientific approach to the design of drugs was Paul Ehrlich, who in 1909 synthesized an organic compound of arsenic which selectively killed the spirochaete that causes syphilis. It was the six-hundred-and-sixth compound that Ehrlich had tested.

During the nineteenth century artificial dyes had been discovered which stained bacteria selectively, and this had led many bacteriologists to look for dyes which would selectively kill bacteria just as they selectively stained them. All the dye substances tested proved to be too toxic for animal cells until in 1935 Gerhard Domagk, a German chemist, found that the dye 'prontosil' could cure experimental infections in mice. Prontosil consists of a dye molecule joined to a molecule of sulfanilamide, and further research showed that it was the sulfanilamide part of the molecule which had the antibacterial activity. Sulfanilamides have since proved to be very useful drugs although they have now been largely superseded by other antibiotics such as penicillin.

The action of sulfanilamide on bacterial cells was not understood until some time after the value of the drug had become known. Bacteria are not killed by the sulfanilamide drugs but rather their growth is inhibited – an effect which can be reversed by another chemical called para-aminobenzoic acid. Bacteria were found to convert para-aminobenzoic acid to folic acid, which is a vitamin essential for their growth. This suggested that sulfanilamide drugs act by interfering with the synthesis of folic acid and by cutting off the supply of it within the cell. Experiment showed that this was the case. Animal cells do not synthesize their own folic acid, but obtain it in food, and so are able to work normally in the presence of the drug. Fortunately, the folic acid entering the animal body in the food cannot be used by the bacteria and so they die a slow death.

Alexander Fleming discovered the existence of penicillin as early as 1929, but it was ten years before the discovery was exploited. He found that colonies of the fungus *Penicillium*, growing as contaminants among bacteria on plates of agar jelly, prevented the growth of the bacteria immediately around them. Fleming proceeded to demonstrate that the fungus produced some chemical in solution which killed certain bacteria yet was not at all harmful to animal cells. Unfortunately he found that this substance, which he called penicillin,

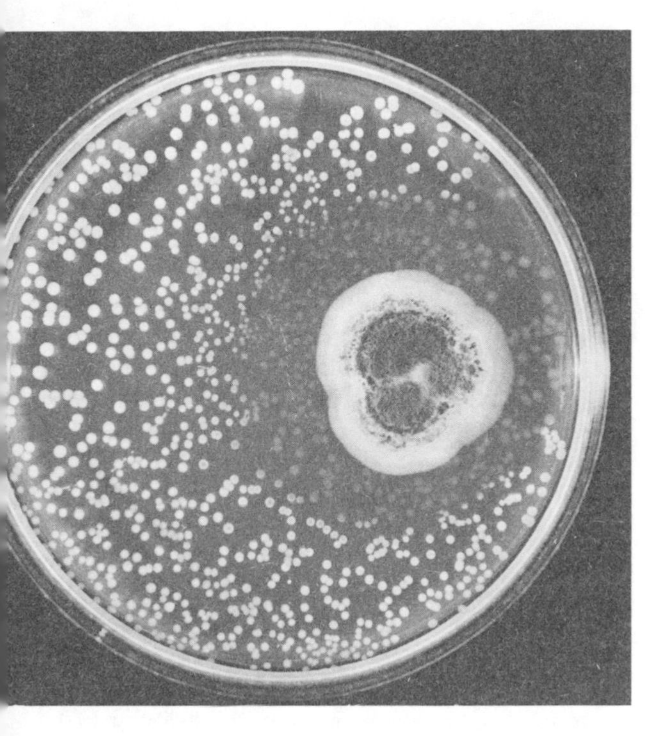

A culture of the fungus Penicillium, from which the antibiotic penicillin is prepared

was very unstable, and, when he found that he was unable to purify it, he abandoned the work. In 1939, however, Sir Howard Florey and Ernst Chain found a way of purifying penicillin without it losing its activity.

Penicillin interferes with the ability of some bacteria to synthesize their cell walls. Animal cells which have no cell walls are none the less able to grow quite normally in the presence of penicillin. It was an extraordinary stroke of luck that penicillin was the first antibiotic to be discovered, since it is still the most generally useful of all the antibiotics. Others have since been found: streptomycin, chloromycetin, aureomycin and terramycin are among the several to have been isolated from actinomycete microbes which grow in the soil. Antibiotics have also been found in plants. Garlic, for example, contains a substance active against certain bacteria. None of these plant antibiotics have so far been found useful in combating human infections.

Organisms resistant to antibiotics have become increasingly common, particularly in hospitals and on farms where antibiotics are used in feeding stuffs. Resistant organisms easily pass from animals to man and from one person to the next. There are various different kinds of resistance to antibiotics. Resistance to penicillin can, for instance, be a result of the production by the organism of an enzyme – penicillinase – which breaks down the penicillin molecule into inactive chemicals. Then, there are other types of penicillin resistance which work by altering the permeability of the cell, so preventing the penicillin from reaching the part of the cell where it would interfere with normal synthesis of bacterial cell wall.

In the 1950s it seemed as though dysentery had been conquered in Japan as a result of the use of sulphonamide drugs; but in 1963 the disease leapt back to its former scourging proportions. Of the bacteria isolated from dysentery patients 10-40 per cent were resistant to all the sulphonamides as well as to streptomycin, chloramphenicol and tetracycline. This new type of bacterial drug resistance has now been found to be caused by a 'drug-resistance factor' which can be passed from one cell to another either during the mating of bacterial cells or by a process of infection. These drug-resistance factors have been found in a number of species of bacteria and it now seems that they can also be passed on from one species of bacteria to another. Unless the use of antibiotics is in future restricted to more urgent medical needs, these resistance factors are likely to prove an ever greater threat as more and more resistant bacteria evolve.

Despite the success in producing anti-bacterial drugs, very little success has been achieved in making anti-viral drugs. At the time of writing there are hopes that two drugs may be useful against viruses, but these drugs have not as yet been tested in full clinical trials. The greatest hope for dealing with viruses still consists in making vaccines that stimulate natural immunity.

There are, of course, many drugs besides those which act directly against viruses and bacteria. One of the most widely used drugs is aspirin, a comparatively simple organic chemical. Originally, a substance related to aspirin was isolated from willow bark, purified and modified chemically to give acetyl-salicylic acid, or aspirin. Aspirin relieves pain, lowers the temperature during fevers and reduces inflammation in rheumatic joints. The way aspirin acts is still not properly understood, although the drug has been in use for seventy years or more. However, some experiments have given results suggesting that aspirin may prevent pain-inducing substances, called kinins, released in damaged tissues from stimulating nerve endings and causing pain.

Morphine, the most useful pain-killing drug, is an extract of opium, which in turn is obtained from the seed pod of the opium poppy. Heroin is chemically related to morphine and can be derived from it; so are codeine, pethidine and methadone – all drugs which can be more or less addictive. De Quincey was one of the first men in the West to give a full description of the addictive effects of opium, which he took in the form of laudanum, an alcoholic extract of opium diluted with water. His book *Confessions of an English Opium Eater* was published in 1821. Morphine was used a great deal more in the nineteenth century than it is now. It was commonly recommended for the treatment of fevers, stomach complaints and many other conditions for which better remedies are now available.

The alkaloid drugs morphine, quinine, cocaine, caffeine and nicotine are derived from plants. Not all plants produce such alkaloids – which seem to perform no essential biochemical function in the cell. Nicotine is produced in the root of the tobacco plant and later flows up in the sap to the leaves. The leafy top of a tomato plant, which does not produce any alkaloids, can be grafted on to the root of a tobacco plant. Yet the nicotine that flows into the leafy tomato top bathes the cells without influencing their growth or development in any way. Even more surprisingly the leafy top of the tobacco plant, if grafted on to the tomato root so that it never receives any nicotine in its leaves, still grows and develops normally. This suggests that nicotine and possibly other alkaloids are present not to perform some vital biochemical role but perhaps to protect the plant from animals which might otherwise eat its leaves. One interesting example of this type of protection occurs in a species of wild tomato known to contain an alkaloid which protects the plant against attack from fungus.

LSD and mescalin – the hallucinogenic or psychedelic drugs – are also alkaloids. LSD is synthesized from the alkaloid ergot, which is derived from a fungus infecting rye; mescalin is obtained from the peyote cactus used by the Aztecs and other Indians in religious ceremonies. Hallucinogenic drugs as their name suggests produce changes in visual perceptions and may induce hallucinations. Some people have found these experiences beautiful and elevating, others have been terrified. LSD is not addictive, but psychological dependence on it can develop. Both LSD and mescalin have a chemical structure similar to adrenalin, the hormone produced by the adrenal gland. Adrenalin is injected into the blood from this gland when an animal is frightened or when it responds suddenly to an unusual stimulus. It causes the heart to beat faster, the blood vessels in the muscles to dilate and many other less evident changes to take place as a result of stimulating certain parts of the nervous system. Psychedelic drugs may act by exciting these parts of the nervous system in a similar but abnormal way.

The remarkable effect of psychedelic drugs on the mind has suggested that similar substances, produced naturally in the body, might cause certain types of mental illness such as schizophrenia. It has been suggested that individuals who are schizophrenic may have inherited a gene which upsets their personal biochemistry and causes the permanent or intermittent production of a substance with a psychedelic effect. This hypothesis suggests that the causes of schizophrenia and possibly of other mental illnesses may be similar to those of diseases such as phenylketonuria or porphyria. Several substances normally found in the body might be expected to have a psychedelic effect if present in large enough quantities. However, a great deal of research has failed to show that any of these substances are found in larger quantities in the urine of schizophrenics than in the urine of other people. Furthermore there is no clear evidence that schizophrenia is caused by a single gene or even that it is a disease with clearly defined symptoms. Other suggestions have been made that the schizophrenic condition is a result of behaviour patterns learnt as a child and reinforced by later experience. It may turn out that both theories are partly right, but it is an indication of our ignorance of mental illness that we cannot yet be certain whether the causes lie in the cells or the environment.

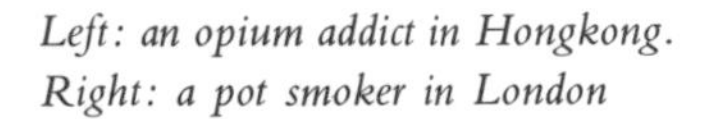

Left: an opium addict in Hongkong. Right: a pot smoker in London

The way in which many drugs act on cells is still not properly understood. This is because the value of most drugs is first recognized as a result of the effect they have on the organism as a whole. These effects are usually the first to be studied and it is left to later research to reveal the fundamental action of the drug on the cells themselves. One of the effects of nicotine is to inhibit the transmission of nerve impulses across the synapses between nerve cells. Many other drugs affecting the nervous system – including the nerve gases – act in similar ways by interfering with the 'transmitter' acetylcholine which carries messages across synapses or with the enzyme acetylcholinesterase which breaks down the acetylcholine after the message has been transmitted. There are other transmitter substances in the brain which are susceptible to drugs in an analogous way.

There is evidence that morphine is a drug that acts on the nervous system by altering the transmission of impulses from one nerve to another. Addiction to morphine clearly has some basis in the cells of the brain itself. Morphine addicts crave larger and larger doses to obtain the same effects on the body, and the body cells become so accustomed to the drug that they can no longer function normally without it. Evidence for this comes from the observation that the respiration of slices of normal rat brain is depressed by morphine, whereas the respiration of slices of rat brain taken from morphine tolerant rats is not depressed by morphine.

Cells against disease

Individual cells pursue a direct battle with viruses and bacteria – at the crudest level it is a question of eat or be eaten. If the cell can trap a virus or bacterium in a food vacuole, it can then digest it with enzymes. Many protozoan cells feed in this way and the phagocytes of multicellular organisms do much the same; by their mopping up activities they protect the body to which they belong. However, vertebrates, and other multicellular animals, have other more elaborate means of defence as well.

Work on the immune defences of the blood got under way in the 1890s, when it was shown that blood serum from immunized animals contained active ingredients which could destroy or agglutinate bacterial cells and precipitate bacterial toxins. These reactions were always found to be specific: in other words, the blood serum was always found only to agglutinate the type of bacteria which had previously been used to immunize the animal. The active ingredients in the blood which do this are the antibodies and the specific parts of the cells with which they react are the antigens.

A wide variety of foreign substances will act as antigens if injected into the blood of animals; proteins and carbohydrates are particularly antigenic. Under normal circumstances the body does not produce antibodies against its own tissues. However, the body does make 'natural antibodies', which are possibly formed as protection against substances we eat, but which also react against foreign transfused blood. Blood used for transfusion has to be carefully crossmatched to ensure that the blood of donor and recipient are compatible and that the recipient does not have natural antibodies to the blood of the donor.

Antibodies are proteins found in the blood serum – the clear substance which remains after the rest of the blood has clotted. If the body of an animal is stimulated it will make individual antibodies against hundreds or even thousands of different antigens. Each of these antibodies is different and will react specifically with its own antigen. How cells manage to produce such a wide variety of individual antibodies is the subject of intense research and some answers are now beginning to emerge.

The antibodies are a complex mixture of protein molecules – the α-, β-, and γ-globulins, which can easily be separated from one another, but each of which is nevertheless a complex mixture. Work on γ-globulin shows that it is a large protein molecule with a small carbohydrate fraction. The protein molecule consists of four polypeptide chains – two long heavy

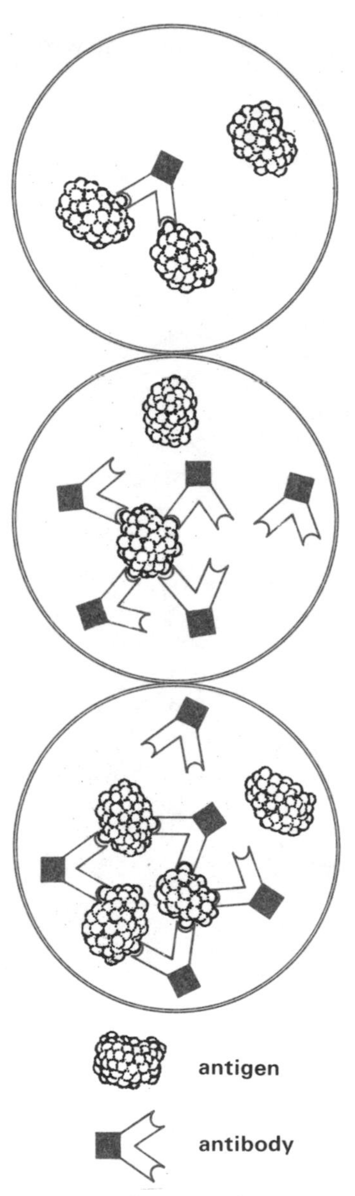

Diagrams illustrating how antibodies circulating in the blood are able to combine with antigens, inactivating them and finally destroying them

chains and two short light chains. This molecule can be split up by protein-digesting enzymes into two or three portions. One portion, called the F*c* fragment, can be crystallized and individual molecules of it are always identical. The other portion, called the antigen-binding fragment F*ab*, consists of two identical halves, which have been found to bind the antigen. The F*ab* fragment varies in structure and cannot be crystallized.

These antibody molecules can be seen under the electron microscope as the result of a very elegant technique devised by scientists working at the National Institute for Medical Research in London. An artificial antigen was synthesized which consisted of two dinitrophenyl molecules separated by a chain of eight carbon atoms. The antigen was injected into a rabbit which, after several days, produced antibody to it. The rabbit antibody was then mixed with the dinitrophenyl antigen in a test tube and the solution examined with an electron microscope. Antibody molecules could be seen associated together in ring-like or triangular structures. These rings consisted of three molecules of antibody joined together by antigen. The combination of three molecules was just large enough to be seen with the electron microscope; individual molecules could not be seen by themselves. These experiments showed – what had long been suspected but never before seen – that some antibodies have more than one binding site with which they can attach to antigens. The experiment was repeated using antibody which had been split by enzymes. Using only the F*ab* portion of the antibody the corners of the triangle were found to be missing, showing that the F*c* portion of the molecule was not concerned with the binding of the antigen.

So many different antibody molecules are made in the animal body that it might seem to be an impossible task to be able to purify one type and obtain enough of it for a full analysis. But here the misfortune of people who suffer from a rare cancer of the white blood cells, called myelomatosis, has benefited science. In these people a great deal of one kind of antibody – the myeloma protein – is produced, and it can be obtained in sufficient quantity for analysis. The myeloma protein taken from a particular person is always the same in composition, but myeloma proteins from different people vary a great deal in composition. However, the F*c* fragment, obtained by enzyme digestion, is always constant in composition and all the variation between different myeloma proteins occurs in the F*ab* fragment.

More detailed analysis has shown that the portion of the F*ab* fragment next to the F*c* fragment is also invariable and that all the variation in the molecule occurs towards the end of the F*ab* fragment. As many as 40 amino-acid sites in the variable portion may actually vary and one of at least five different amino acids may be found at each site – which means that a very large number of different molecules is possible. This variability of the antibody molecule is the means by which antibodies can be made to match any antigen. How the cell contains enough genetic information to synthesize such almost infinitely variable molecules is a research problem currently receiving a great deal of attention. It seems possible that the wide variety of molecules may be produced as the result of some process of directed mutation of a part of the genetic material – a sort of scrambling of the gene, the variable portion of the molecule being specified by a gene formed by a process of repeated breakage and repair. Individual cells in the spleen and lymph nodes appear to specialize in producing one particular antibody. When an antigen which fits the antibody produced by a particular cell enters the body it stimulates that particular cell to divide and give rise to many descendants, which between them produce a great deal of the specific antibody. According to this theory, for which there is increasing evidence, an organism becomes immune as a result of a great multiplication in the number of cells which produce antibody to a particular antigen.

Certain parts of the body such as the lens of the eye, nervous tissue and spermatozoa are not in direct

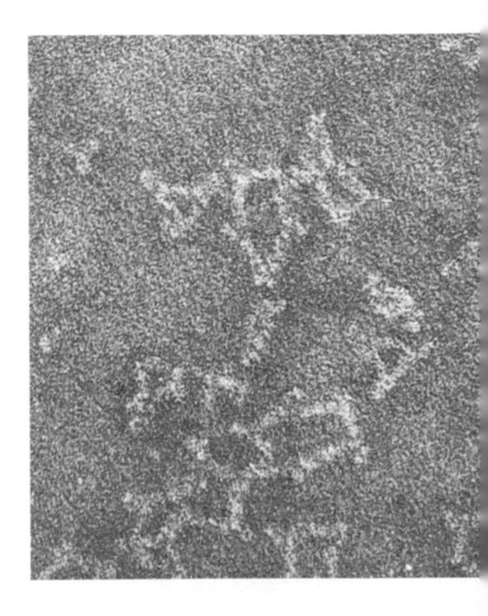

Above and below: electron micrographs of antibody particles magnified 400,000 times

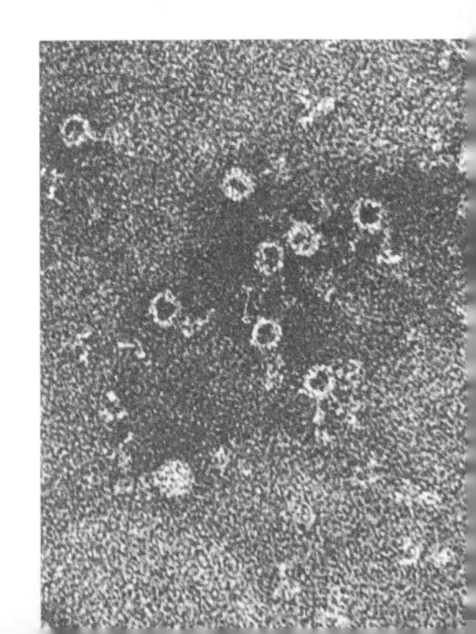

contact with blood. If portions of these parts of the body are removed in experimental animals and injected into the blood then the body will form antibodies against them. These are called auto-immune antibodies and occasionally may be produced by accident in man when damaged tissues get into the blood. Auto-immune reaction to sperm causes sterility and auto-immune reaction to nervous tissues leads to severe brain damage. However, the body does not normally make antibodies against itself – there seems to be some sort of process of self-recognition. In newly born animals the process of self-recognition has not been completed. Animals injected with foreign proteins at this age become tolerant to them so that they cannot make antibodies to such proteins in later life. In fact there is a danger of their becoming tolerant to mildly infectious viruses which may reappear to cause cancers in the defenceless body at a later age. The tolerance to proteins, which are present in the body at this early age, is believed to occur as a result of the actual elimination of cells in the lymph glands which would be capable of making antibody to the proteins. How this happens is not known. Within about a week after birth the situation is stabilized and any new foreign material entering the body after this age will produce an immune reaction.

The body has other defences besides antibodies. The phagocytes work in conjunction with the antibodies and eat up foreign matter. There is also a battery of enzymes, called complement, which destroys foreign cells in the body by punching holes in their outer membranes so that the cytoplasm leaks out. The cells of higher animals have also been found to have a special means of protection against infection by viruses. During the course of infection by a virus, a substance is produced, called interferon, which protects uninfected cells from infection. Interferon can be obtained from animals or animal cells after they have been infected with a virus. Interferon has been purified and is now known to be a protein, but it is quite different from antibody. It can be produced by all the cells of the body, not just special cells. Interferon from monkeys has been shown to give protection to man in experimental infections with cowpox – the virus which is used to vaccinate against smallpox. However, interferon taken from one species of animal does not in most cases give any protection against virus infections in other species and this will probably be a severe restriction to its usefulness as a drug. The synthesis of interferon is stimulated by viral nucleic acid and it is made in the same way as the other protein of the cell. The way in which interferon acts is not completely understood but it seems to do so by preventing the virus from multiplying within the cell, rather than by preventing the virus from actually entering the cell.

Do cells grow old?

The cells of micro-organisms are in a sense immortal. A single cell of a bacterium can divide to give thousands or millions of identical copies of itself. So long as some of the daughter cells go on living and dividing and no mutations occur it is possible to say that in a sense the original cell is still alive. The bodies of multicellular animals and plants do not, however, live forever, although trees do live for thousands of years. It has been calculated, for example, that a number of thriving giant sequoia trees in California are well over two thousand years old.

There are many cultivated plants which are propagated entirely by taking cuttings. If healthy growing cells or tissues are chosen for propagation by cuttings, the daughter plants never show any effects of aging. On the other hand, annual plants normally die after flowering and live for less than a year. If, however, the flowers are removed and flowering is prevented, annuals may live for two or more years. The flowers of many orchids will remain in perfect condition for weeks or months, so long as they are not pollinated, but begin to wilt as soon as pollen is placed on the female organ. In these cases cell death seems to be programmed as part of the developmental process, in a

Dwarfing a tourist couple are the massive trunks of giant Californian sequoias. Some of these trees are over two thousand years old

way reminiscent of the cell death which occurs in the tail of a developing tadpole.

Different species of animals live for very different lengths of time. A mouse is old at three years, a cat or dog at 12; and people are old at 60 or 70. However, the cells from certain parts of an old animal, such as the gut, are still capable of rapid and healthy growth at the time of death. Pieces of skin can be taken from the ear of a mouse aged one year and transplanted to another young mouse. After another year the same piece of skin can be transplanted once more and so on. The ear skin retains its peculiar character on transplantation and so can be easily recognized. Pieces of skin transplanted in this way have survived for more than six years or twice the normal lifetime of the mouse. However, ovaries from an old mouse transplanted to a young mouse do not function normally, suggesting that ovaries, unlike skin, do age in some way. Cells from animals growing in tissue culture appear to be immortal, like the cells of micro-organisms, but these may not be normal cells. There is some evidence that normal animal cells may die after about 50 cell divisions in tissue culture, leaving only malignant cells, which have mutated, to carry on growing. These ideas are still being disputed, but it seems clear from the grafting experiments in mice that animal cells can go on growing normally for very much longer than their 'natural' span.

It seems most likely that dividing cells do not age at all rapidly and that the aging which is immediately responsible for an animal's death occurs primarily in the non-dividing cells such as the ovary or the brain. Cells die in these tissues without being replaced and so eventually some tissues may not have enough cells to function properly. The death of the cells of one vital tissue might then cause the death of the whole organism. But this explanation of death is almost certainly an over-simplification, and the full answer must await further experimental discoveries.

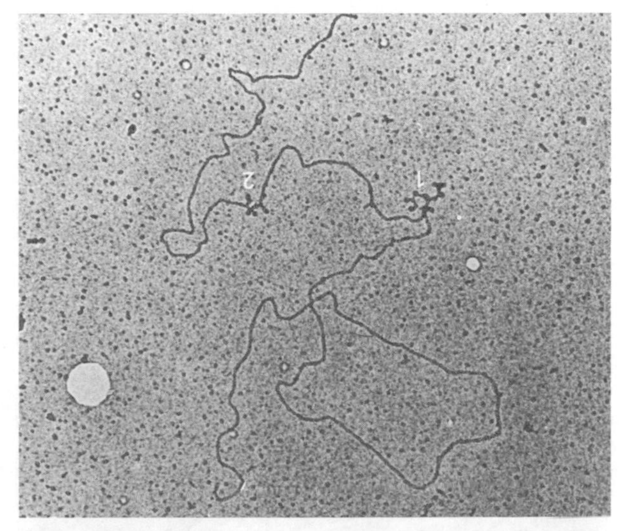

An electron micrograph of a pure gene isolated from the chromosome of the colon bacillus. The development of techniques of isolating individual genes in this way may be bringing nearer the time when the selective deletion or addition of genes in living egg cells will be possible

For many, the prospect of a world in which 'genetic engineering' is a possibility and the difficulties surrounding organ transplantation are finally resolved is an exciting one. For others, it is the prospect of a world in which the horrific creations of a Dr Frankenstein are no longer impossible fictions

Overleaf: predictions of the progress of biological science by a large team of experts. In each case the peak of the left-hand figure indicates the year most popularly predicted as that by which a stated event will have occurred, assuming an even chance. The right-hand peak gives the year most popularly predicted, assuming a 90 per cent chance. The vertical edges of the figures show the time limits, as agreed by 50 per cent of the experts, within which the predicted events will take place. Some events – those with figures crossing the right-hand limit of the chart – may, in the opinion of most experts, never occur ►

Identification of new enzyme systems involved in health and disease

Development of drugs to cure anxiety and tension states

Development of drugs to cure depression

Development of cigarette not liable to cause cancer

Detailed correlation of chromosomal abnormalities with disease

Development of useful tissue adhesives to replace surgical sutures

Development of drugs to control fertility in males

Development of drugs to prevent and cure tooth decay

Rapid, specific diagnosis of diseases caused by viruses

Development of satisfactory method for preserving transplant organs

Transmission of genetic information by using viruses

Development of a clinically useful mechanical heart

Development of drugs for the cure and prevention of drug dependence

Synthesis of a living virus in the laboratory

Development of effective agents for localized, cell-specific cancer therapy

Genetic modification of the developing embryo

Development of electronic devices to cure schizophrenia

Instigation of compulsory human eugenic programmes

Electronic control of human behaviour

Total control of mental development

Artificial synthesis of a living organism

Complete control of human behaviour by chemical means

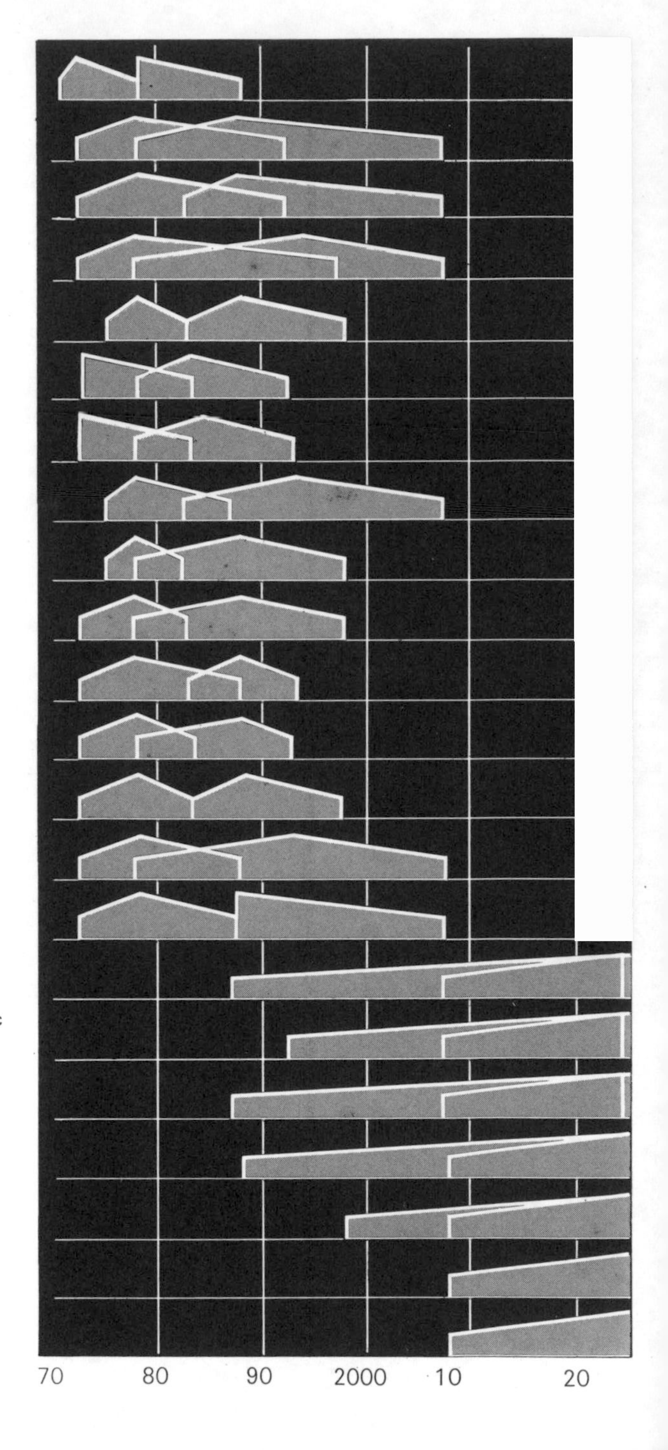

GLOSSARY

amino acid: organic acid containing both amino (NH_2) and acidic carboxyl (COOH) groups. There are twenty different amino acids commonly found in living cells as the constituents of proteins.

antibiotic: chemical substance produced by moulds or bacteria which is capable of selectively killing, or of preventing the growth of, other cell life. Penicillin, which attacks many bacterial species, is the best-known example.

antibody: protein molecule produced in large quantities by the animal body in response to the presence, in the body, of a foreign substance – an antigen. It is the front line of defence against disease.

antigen: a foreign substance in the body which stimulates the production of antibody molecules.

atom: the smallest quantity of a chemical element which can combine with another chemical element to form a chemical compound or molecule.

biochemical pathway: route by which substances pass from one enzyme reaction to the next, changing from one compound to another as they progress. These pathways do not exist as pieces of cytoplasm but can be determined by noting the itinerary of radioactive tracer molecules.

blastula: hollow ball of cells produced by the division of an egg cell.

carbohydrate: substances such as starch, sugars and cellulose which consist of carbon, hydrogen and oxygen only.

centriole: granule which doubles just before cell division in higher plants and seems to be important for the normal formation of the spindle.

centrosome: region of cytoplasm containing one or two centrioles.

chlorophyll: green pigment of plants which plays a vital role in photosynthesis.

chloroplast: small body in the cytoplasm of plant cells which contains the chlorophyll.

chromosome: thread of DNA (or, in certain viruses, RNA) which carries the genetic information. In eucaryotes it is covered with a protective coat of histone protein.

cilia: fine threads of cytoplasm which project from the cell and beat together with oarlike movements to push the surrounding liquid past the cell or propel a free cell through the liquid.

cytoplasm: all the substance of the cell excluding the nucleus.

DNA: deoxyribonucleic acid – the genetic material of all cells and many viruses. It is a polymer of four bases – adenine, guanine, thymine and cytosine – which are joined end to end to form a double helix. The bases are paired across from one strand to the other of the double helix, adenine always pairing with thymine and guanine always with cytosine.

endoplasmic reticulum: rough and smooth membranes in the cell, arranged like the skins of an onion. The rough membranes are dotted with ribosomes and are actively synthesizing proteins; the smooth membranes are inactive.

enzyme: protein substance which speeds up, or catalyses, the chemical reactions of the cell.

eucaryote: plant or animal cell with true nucleus surrounded by a nuclear membrane and having chromosomes covered in histone protein.

fat: substance consisting only of carbon, hydrogen and oxygen and containing organic acids.

feedback control: control of the flow of substances through a biochemical pathway by the action of the product of the pathway. The product may either inhibit the action of an enzyme 'earlier' in the pathway or affect the rate of synthesis of the enzyme.

flagellum: fine long thread of cytoplasm used to propel the organism through liquid by thrashing movements. Found on flagellate protozoa and forming the sperm tail.

gastrula: stage in the development of the embryo following the blastula.

gene: originally defined as an abstract concept deduced from experiments in breeding, it is now known to be a portion of DNA devoted to carrying the code for a particular protein.

golgi apparatus: network of small vacuoles in which potentially harmful proteins, like digestive enzymes, are packaged for transport through the cell cytoplasm before delivery outside the cell.

haemoglobin: the red pigment of the blood which carries oxygen around the body.

histone: protein covering the chromosomes of eucaryotic cells. It protects the chromosomes and may also regulate their activity.

hormone: substance produced in small quantities in one part of the body which controls the functions of other parts of the body.

inorganic chemical: chemical originally defined as being of mineral origin; not an organic chemical.

lysosome: small body present in large numbers in the cytoplasm, containing powerful enzymes capable of digesting the cell or other organic material. Released into digestive vacuoles for this purpose.

membrane: organized layers of protein and fat. For example: the outer surface of the cell, the endoplasmic reticulum and the nuclear membrane.

mesosome: structure, found only in the bacterial cell, which contains the respiratory enzymes and may also be of importance in the division of the cell.

messenger RNA: ribonucleic acid molecules which carry the genetic message from the nucleus to the cytoplasm and act as a template on which the proteins are manufactured.

metabolism: the chemical processes of living organisms.

mitochondrion: small rod-shaped body, usually present in large numbers in the cell, concerned with supplying the cell with energy derived by the aerobic breakdown of sugars.

molecule: smallest portion of a substance capable of existing independently.

mutation: sudden change in the genetic material, which can be caused, for example, by a mistaken copying of the genetic message.

nucleus: the part of the cell which contains the chromosome(s).

nucleolus: small body or bodies consisting mostly of RNA attached to a particular place on one or more chromosomes.

organelle: any small body, such as a mitochondrion, lysosome, chloroplast, cilium etc., which is part of the cytoplasm of the cell.

organic chemical: originally a chemical substance derived from living material. Now these chemicals can very often by synthesized in the test tube and the term has come to mean any chemical compound containing carbon.

photosynthesis: synthesis of chemical, such as sugars, by using energy from sunlight.

polymer: substance, such as cellulose, polythene, DNA, RNA and proteins, consisting of a number of subunits joined together to form a chain.

polysome: number of ribosomes attached to a molecule of messenger RNA.

procaryote: organisms, such as bacteria, which lack a true nucleus, histone protein and cellular organelles.

protein: molecules made up from a number of different amino acids arranged in a special order determined by the genetic code. Enzymes and also many structural parts of the cell consist of protein.

protoplasm: the substance of living cells.

reflex: a simple response to a stimulus, resulting from the passage of messages through the nervous system; for example, the rapid withdrawal of the hand when it is pricked.

repressor substance: substance which 'switches off' the genetic material; shown in at least one case to be a protein.

ribosome: small particle which consists of protein and RNA and plays a vital role in protein synthesis. Most cells contain many thousands of ribosomes.

RNA: ribonucleic acid is important as the messenger RNA of the cell and the transfer RNA and is also an essential part of the ribsome, too. It also forms the genetic material of certain viruses.

synapse: a junction of fibres coming from two different nerve cells.

transfer RNA: RNA molecules which can attach themselves to a particular part of the messenger RNA at one end and to a specific amino acid at the other end, thus bringing the amino acids together in the right order before they join up to form a protein.

FURTHER READING

Alexander, Peter, *Atomic Radiation and Life*. Harmondsworth, 1962.

Andrews, Sir Christopher, *The Natural History of Viruses*. London & New York, 1967.

Anfinsen, Christian B., *The Molecular Basis of Evolution*. London & New York, 1959.

Asimov, Isaac, *The Chemicals of Life*. New York & London, 1958.

—*The Genetic Code*. New York, 1963; London, 1964.

Beadle, George and Muriel, *The Language of Life*. London, 1968.

Butler, J. A. V., *The Life of the Cell*. London, 1966.

Clowes, Royston, *The Structure of Life*. London, 1967.

Crick, Francis, *Of Molecules and Men*. Washington, 1967.

Dobzhansky, Th., *Evolution, Genetics and Man*. London & New York, 1955.

Galamtos, Robert, *Nerves and Muscles*. London, 1965.

Harris, R. J. C., *Cancer*. Harmondsworth, 1965.

Hutchins, C. M., *Life's Key – DNA*. New York, 1961.

Jevons, R. F., *The Biochemical Approach to Life*. London & New York, 1964.

Kendrew, J. C., *The Thread of Life*. London, 1966.

The Living Cell. Readings from *Scientific American*. San Francisco & London, 1965.

Medawar, Sir Peter, *The Future of Man*. London & New York, 1960.

The Molecular Basis of Life. Readings from *Scientific American*. San Francisco & London, 1965.

Oparin, A. I., *Life: Its Nature, Origin and Development*. New York, 1962.

Rattray Taylor, G., *The Science of Life*. London, 1963.

—*The Biological Time Bomb*. London, 1968.

Rose, Steven, *The Chemistry of Life*. London, 1966.

Rosland, Jean, *Can Man be Modified?* London, 1959.

Schrödinger, Erwin, *What is Life?* Cambridge, 1944.

Smith, Kenneth M., *Beyond the Microscope*. London, 1957.

—*Biology of Viruses*. Oxford, 1965.

Stanier, R. Y., Doudoroff, M., and Adelberg, E., *The Microbial World*. New York, 1965.

Stanley, Wendell & Velens, E. G., *Viruses and the Nature of Life*. New York, 1961; London, 1962.

Watson, J. D. *The Double Helix*. London & New York, 1968.

Wedberg, S. E., *Introduction to Microbiology*. London & New York, 1966.

ILLUSTRATION SOURCES

Page 8 Frog. Bettina Gruber.

9 Moss agate. Courtesy of Josephine Marquand; Oliver and Boyd.

10 Candle flame. Courtesy of Carl Zeiss.

Growth spiral on silicon carbide crystal. Courtesy of S. Tolansky.

11 Amoeba. Gordon F. Leedale.

12 Will Burtin's model of the cell. Courtesy of the Upjohn Company.

13 'The New Eve'. Cartoon by 'Serre'. *Science et Vie*, October 1969.

14 Descent of human soul into embryo. Nassauische Landesbibliothek.

16 Metabolic pathways. D. E. Nicholson.

Electron micrograph of cells of onion root tip. Rockefeller Institute.

17 Mesophyll leaf cells M. I. Walker.

18 Cellulose lamellae in walls of *Chaetomorpha melagonium*. Eva Frei and R. D. Preston.

20 Model of part of the DNA molecule. R. E. Barker.

21 Fossil bacteria from Silurian saltbeds. Courtesy Editor of Bild der Wissenschaft and H. Dombrowski.

23 Stanley Miller and his 'origin of life' experiment. Courtesy Stanley L. Miller.

24 Sydney Fox's protenoid microspheres. Electron micrograph. Steven M. Brooke.

25 Sonorous figure. Plate: dia. 32cm, thickness 0.55 mm; the material is strewn quartz sand; frequency 8200cps. From Hans Jenny's book *Cymatics*. Basilius Press, Basle.

27 Genesis. An illustration from the Cambridge Bible of 1663. The Mansell Collection.

28 Transverse section through annual ring of Scotch Pine. Harris Biological Supplies.

29 Cork cells. Robert Hooke: *Micrographia*, 1665.

30 M. J. Schleiden and T. Schwann. W. Stirling, *Some Apostles of Physiology*, 1902.

T. Schwann. Microscopical Researches 1847.

31 Hooke's own microscope by Christopher Cock c1675. Science Museum, London.

32 Super-speed 65 ultra-centrifuge. Courtesy of Measuring and Scientific Equipment Ltd, London.

33 Electron microscope. AEI Scientific Apparatus Ltd.

34 Animal cell. John D. Dodge.

35 Plant cell. Gordon F. Leedale.

36 Nucleus of the dinoflagellate *Amphidinium*. John D. Dodge.

37 Enlarged section through nucleus of dinoflagellate. John D. Dodge.

38 Salivary gland chromosomes of *Smittia parthenogenetica*. Professor H. Bauer.

40 Bar-eye syndrome in Drosophila. Brookhaven National Laboratory.

41 Chromosome from salivary gland cells of *Chironomus tentans*. W. Beerman.

45 Golgi body in *Khawkinea*. Gordon F. Leedale.

47 Monkey kidney cells. M. R. Young. National Institute for Medical Research.

50 Longitudinal section through mitochondrion of euglena. Gordon F. Leedale.

54 Chloroplast of euglena. Gordon F. Leedale.

56 Whole flagellum of *Bumilleria*. Gordon F. Leedale.

57 Transverse section through cilia. Bjorn Afzelius. *Journal of Ultrastructure Research*, Academic Press.

63 Meiosis in *Trillium erectum*. Brookhaven National Laboratory.

64 Section through maturing spermatids in the human testis. T. Nagano, *The World Through the Electron Microscope*; and Springer-Verlag.

Human ovum containing two pronuclei. Zeev Dickman.

67 *Euglena proxima*. Gordon F. Leedale.

68 Moravian monk, Gregor Mendel. Radio Times Hulton Picture Library.

73 Chromosomes. Patricia A. Jacobs.

83 Lung X-ray. J. Harrison. Luton and Dunstable Hospital.

Iris leaf stomata. Gene Cox.

84 John Whetton winning the one mile. E. D. Lacey.

86 François Jacob. Institut Pasteur.

Jacques Monod. Institut Pasteur.

92 E. coli bacteriophage (T2) releasing DNA molecule after osmotic shock. A. K. Kleinschmidt, New York University Medical Center. Courtesy Elsevier Publishing Co.

97 X-ray diffraction picture of DNA. Maurice Wilkins.

98 Francis Crick and James Watson. Courtesy of Weidenfeld & Nicholson. Photograph A. C. Barrington Brown.

100 DNA replication. Drawing by Gordon Cramp.

101 John Kendrew, C.B.E., Sc.D., F.R.S. Edward Leigh.

102 Model to show folding of myoglobin molecule. Sunday Times photo by Ian Yeomans.

Electron-density distribution map of insulin, drawn by Harald Hager. In H. W. Franke, *Microstructures of Chemistry*, Basilius Press, Basle.

103 Contour map of myoglobin. J. C. Kendrew.

104 Normal human blood cells. A. C. Allison.

105 Sickled human blood cells. A. C. Allison.

108 Polyribosome from a tobacco leaf. R. G. Milne. Rothamsted Experimental Station.

115 Antony van Leeuwenhoek (1632-1723). From the original mezzotint of 1686 by J. Verkolje. Impression in the Wellcome Institute of the History of Medicine. By courtesy of the Wellcome Trustees.

116 *Vorticella.* M. I. Walker.

117 Leeuwenhoek's microscope. The Science Museum, South Kensington.

120 *Pediastrum.* M. I. Walker.

121 *Stentor.* M. I. Walker.

Paramecium bursaria M. I. Walker.

122 *Amoeba.* M. I. Walker.

123 Radiolaria, drawn by Haeckel for his *Report on Radiolaria,* 1887.

124 Foraminifera. M. I. Walker.

125 Radiolaria. M. I. Walker.

127 *Euglena spirogyra.* Gordon F. Leedale.

128 *Volvox aureus.* M. I. Walker.

129 *Euglena ehrenbergii.* M. I. Walker.

131 Hydra. Flatters and Garnett Ltd, Manchester.

132 Diatoms. M. I. Walker.

133 Diatom. M. I. Walker.

135 *Bacillus proteus* photographed with the 1MV electron microscope (contrast-stop method) Laboratoire d'Optique Electronique du C.N.R.S., Toulouse.

Clostridium botulinium. J. Harrison, Luton and Dunstable Hospital.

137 Living bacteria. M. I. Walker.

138 Electron micrograph of lecithin/cholesterol mixture negatively stained. R. W. Horne.

140 Sowbane mosaic virus particles. R. G. Milne, Rothamsted Experimental Station.

141 SP-50 phage on *Bacillus subtilus.* R. G. Milne. Rothamsted Experimental Station.

Influenza virus. H. G. Pereira. Courtesy British Medical Bulletin.

142 Tobacco mosaic virus. R. G. Milne. Rothamsted Experimental Station.

Adenovirus. R.W. Valentine.

Model of Adenovirus. R. W. Horne.

143 T2 'triggered' phage. R. W. Horne, S. Brenner. Courtesy Journal of Molecular Biology.

T2 bacteriophage adsorbed on empty cell walls of *E. coli.* E. Kellenberger. Laboratoire de Biophysique de l'Institut de Biologie Moleculaire. University of Geneva. Z. Naturforschg 10b 698–704 (1955).

Part of a lysed cell. Phage T4. E. Kellenberger, G. Kellenberger. Courtesy of *Scientific American.*

144 Human embryo head. Erich Blechschmidt.

147 Cleavage of the frog egg. Arnaldo Legname.

150 Hybrid cell containing a human nucleus and a mouse nucleus. H. Harris.

Chromosomes of man/mouse hybrid cell mitosis. H. Harris.

151 Colony of mononucleate man/mouse cells. H. Harris.

154 Transplantation of nucleus of frog blastula cell. J. B. Gurdon.

155 Genetically identical cloned frogs. J. B. Gurdon and the Journal of Heredity.

158 Human embryo, showing development of the eye. Erich Blechschmidt.

159 Human embryo head. Erich Blechschmidt.

160 Enlarged and normal kidney. J. Harrison. Luton and Dunstable Hospital.

164 Mongol child. WHO picture by D. Henrioud.

166 Thalidomide child. World Medicine.

168 Striated muscle from rabbit. H. E. Huxley

169 Portion of cerebral cortex stained by Golgi method. J. Z. Young. Science Journal.

Nerve cell. J. Z. Young.

170 Luigi Galvani demonstrating the electrical nature of muscle action. Universita degle Studi, Bologna.

173 Schleiden's sketch of embryonic material in flowering plants. M. J. Schleiden, *Beitrage zur Phytogenesis,* 1838.

175 Section through stem of lime (Tilia). Gene Cox.

Fungal reproductive body. John D. Dodge.

178 Cancer cell. Chester Beatty Research Institute and Cambridge Instrument Company.

180 Basal cell epithelium. Institute of Dermatology, St. John's Hospital.

182 Underwater atomic explosion. USIS.

183 Victim of Hiroshima showing radiation burns. Keystone Press.

184 Mangled chromosomes – a result of radiation. Brookhaven National Laboratory.

185 Normal and cancerous lung tissue. J. Harrison. Luton and Dunstable Hospital.

Cancerous lung tissue. J. Harrison, Luton and Dunstable Hospital.

186 Bone cancer in adult male. Thames and Hudson archives.

Bone cancer in adult male. Thames and Hudson archives.

187 Radiograph of brain tumour. Institute of Neurology, the National Hospital.

189 James I; and George III. The Mansell Collection.

193 Reconstruction of Fleming's discovery of bacteriocidal effect of Penicillin mould. I.C.I.

197 Opium smoker. WHO photo.

Pot smoking. World Medicine.

201 Antibody molecules. N. M. Green.

203 Sequoia trees in Yosemite National Park. USIS.

205 Isolated gene. Science Journal.

Frankenstein's monster. Hammer Productions.

The diagrams were drawn by John Stokes.

INDEX